THE SMELL OF FRESH RAIN

THE SMELL OF FRESH RAIN

THE UNEXPECTED PLEASURES OF OUR MOST ELUSIVE SENSE

BARNEY SHAW

First published in the UK in 2017
by Icon Books Ltd, Omnibus Business Centre,
39–41 North Road, London N7 9DP
email: info@iconbooks.com
www.iconbooks.com

This edition published in the UK in 2018 by Icon Books Ltd

Sold in the UK, Europe and Asia
by Faber & Faber Ltd, Bloomsbury House,
74–77 Great Russell Street,
London WC1B 3DA or their agents

Distributed in the UK, Europe and Asia
by Grantham Book Services,
Trent Road, Grantham NG31 7XQ

Distributed in the USA
by Publishers Group West,
1700 Fourth Street, Berkeley, CA 94710

Distributed in Australia and New Zealand
by Allen & Unwin Pty Ltd,
PO Box 8500, 83 Alexander Street,
Crows Nest, NSW 2065

Distributed in South Africa
by Jonathan Ball, Office B4, The District,
41 Sir Lowry Road, Woodstock 7925

Distributed in India by Penguin Books India,
7th Floor, Infinity Tower – C, DLF Cyber City,
Gurgaon 122002, Haryana

Distributed in Canada by Publishers Group Canada,
76 Stafford Street, Unit 300
Toronto, Ontario M6J 2S1

ISBN: 978-178578-341-8

Typeset in MT Garamond by Marie Doherty

Printed and bound in the UK

To Frances

Contents

CHAPTER 1

The Smell of Fresh Rain

> 'In this hall the youth of Market Snodsbury had been eating its daily lunch for a matter of five hundred years, and the flavour lingered. The air was sort of heavy and languorous, if you know what I mean, with the scent of Young England and boiled beef and carrots.'
>
> P.G. Wodehouse – *Right Ho, Jeeves*

Smells are inarticulate. We struggle to understand what they mean. We struggle to put them into words. It is common to experience what psychologists call a tip-of-the-nose effect: we sniff something; we have a fleeting sensation that we almost recognise the smell; we cannot quite pin it down; and then we breathe in a new breath and a new smell. That smell was on the tip of our noses, but, to our frustration, we did not have time to place it. Even if we can recognise it, it is hard to describe it. We do not seem have a readily accessible vocabulary to describe it, as we would if we were seeking to describe a colour or a sound. There is no common language for smells, and we are driven, like P.G. Wodehouse, to vague, hand-flapping descriptions, using phrases such as 'sort of heavy'. Or we resort to feeble tautologies such as 'these capers smell … caperish'. Or we use grand but uninformative words that disguise our inarticulacy like 'indescribable', 'characteristic', 'unmistakable', 'evocative', 'insidious'.

It is not that the sense of smell has only dull things to tell us. It is not like a kick on the shin, which our sense of touch instantly recognises as a kick on the shin, and our memory tells us will get worse before it gets better and there is no more to be said about it. Smells have meanings. They are often charged with emotion. They are often full of nostalgia. They bond us to our nearest and dearest. They often seem to be curiously related to the smell of something quite different, which ought not to smell the same, as if there were connections between things that are hidden and inner meanings that are not obvious to the eye. They sometimes give us a shock of familiarity, or a shock of unfamiliarity. Occasionally a smell will transport us to our distant past – to our first school playground or a warm day in some childhood room – and will bring it alive more vividly and insistently than our other senses can. We often want to pin down what the smell is of, what it is like, what point in our lives it harks back to. The meanings are frustratingly close to hand but difficult to bring to mind.

It is not that smells give us no pleasure. Think of some of the delightful aromas of cooking: of newly baked bread, of grilling on charcoal, of the full-blooded smell of a delicatessen with salamis and cheeses and marinated olives. Think of the smells of cleanliness: of clean linen, of ironing and airing cupboards, of beeswax polish. Think of the scents of Christmas: of mandarin oranges and pine needles and roasting chestnuts. Think of the smells of summer: of beaches and suntan lotion, of mown grass and hay. Think of all the varieties of smoke: the smoke of autumn bonfires, of winter coal, of cigars, of barbecues. Think, should it be your pleasure, of the special smells of very new babies or very old cars, of a timber yard or a building site, of a pub. The variety of delightful smells is almost limitless, and the human nose, however modest we may feel about our olfactory powers, is an instrument of exceptional discrimination.

Many of us are diffident about our ability to detect smells. We might assume that our sense of smell is poor, and that we are defective in this domain. We tend to assume that it takes a special creature – a sniffer dog or a perfume expert – to make sense of smells. We tend to be puzzled by smells and to feel that it is beyond us to understand them properly. But if we do this, we do not give our noses the credit they deserve. Most of us have noses that are really rather brilliant, that can detect minute traces of chemicals that are too tiny to be visible or audible or touchable. Without knowing it, we are using our noses every day to interpret the world around us in the finest detail, and we are unconsciously discriminating between aromas with the greatest subtlety.

I have lived with my sense of smell for years without understanding what it tells me. Instead, I have been puzzled in many different ways. It was a puzzle to me to find a simple smell transporting me 40 or 50 years back in time, instantly and vividly at a single sniff. It was a summer afternoon a couple of years ago. I stepped outside to the smell of fresh rain. A summer shower was falling. Fat rain drops landed on warm tarmac and earth. The gutters were running with water, in which unusually fat bubbles wobbled towards the drain. Rafts of leafy material, the casings of leaves newly shed by lime and plane trees along the street, wavered along with the water and here and there formed temporary dams. The air smelled of rain on newly moistened earth and leaves, a smell which instantly transported me to my childhood.

I was back in that suburban street where my parents had

their house, where 50 years ago the fat raindrops of a summer shower made unusually fat bubbles wobble down a similar gutter towards a similar drain. Not only did the smell remind me of the smell all those years ago, not only did it remind me of the look of that childhood street – the fat bubbles and the grey-green granite kerbstones and the Ford Zephyr parked in the street. It revived also in the same instant an internal memory bringing all the senses into action – the sensation of being seven years old, close to the wavering bubbles, with bare knees and scratchy woollen clothes. Like a seed that lies dormant underground for decades until brought to the surface by a plough, the olfactory memory lies dormant in our heads until a passing whiff makes it germinate. The memories of smells seem to last longer and emerge fresher than memories of sights or sounds, while those senses give us memories that become muddled and faded over time, and do not carry so powerful a charge of recollection and meaning.

There are more puzzles about the sense of smell. I have been puzzled that there are many smells which give a nudge of recognition, but which, unlike the scent of fresh rain on warm earth, cannot be placed. The smells of the street, for instance, waft a succession of changing aromas into my nose as I walk along, aromas which I am on the verge of recognising, which are on the tip of my nose, which might be perfume or privet or catmint or fox or rubbish bin, but are just out of reach. It is as if my sense of smell is not quite alert enough to interpret the scents as they waft into and out of my nose, as if it is always just below the level of conscious attention. It is a puzzle that, although my brain is interested in recognising such smells, and is equipped to do so, it does not know how to speak the language of smells. I want to know whether my sense of smell is simply incapable of speaking that language, or whether it is a matter of practice and effort to do so.

I have been puzzled that there are aromas – many of them – which I can recognise at a sniff, and can successfully place, but I cannot describe. I think, for instance, of bracken: the archetypal smell of damp common land in the British Isles, a fragrance that is strong and distinctive and moving. I can recognise the scent of bracken without hesitation; it makes me think straight away of the fringes of open hills and moors where it grows. But I cannot find the words that fit the smell of bracken and distinguish it from the smell of other green things. More than that, I cannot summon up the smell of bracken in my head. The scent of bracken is more evocative than the sight of it, but whereas I am able to summon up the look of the thing, I am unable to summon up the smell. I can visualise a single frond or a hillside of bracken; I can summon up the feel of running a frond of bracken through my fingers; but the smell is just out of reach.

I have been puzzled too that there are smells which I can recognise and place and (after a fashion) describe, and which have affinities with quite unrelated things. Books, for instance, have their scents which are varied and interesting. The smell of a cheap comic is quite distinct from the smell of a paperback and different again from the smell of a glossy coffee table book. It is no surprise that the smell of these books has a component of damp wood and ink, but why should there also be touches of vanilla, of china clay and of mushrooms? There are, it seems, some hidden meanings beneath the main smell of books. I would like to know how to read those hidden meanings.

Bookish smells, library smells, show up two of the gentle pleasures of the sense of smell. First there is the pleasure of that musty scent, faintly sweet like vanilla, faintly grassy, an occasional whiff of leather and of earth. It is familiar, faded and evocative; small wonder then that you can buy an *eau de cologne* called '*Paperback*'. Second, there is the pleasure that comes because the library smell is telling us hidden facts about the chemistry of books. The bookish scent is the product of several hundred chemical compounds given off in tiny quantities by the paper, the ink, the glue and the binding of books. The compound lignin, which is present in all wood-based papers, is closely related to vanillin, the synthetic flavouring that took over from expensive vanilla imported from such exotic places as the Comoro Islands and Madagascar. As it breaks down, the lignin gives old books that faint vanilla scent. Compounds deriving from pine tar are used to make paper more impermeable to ink, and contribute to the musty, camphorous smell of books. An earthy mushroom odour is caused by other fragrant molecules related to alcohols present in the paper. Another earthy smell comes from the china clay used to give gloss to glossy papers.

Then there are the puzzles of flavour. A runny cheese is one of the most delicious flavours and one of the foulest smells. It smells like damp socks in a gym locker, but it tastes wonderful. However, taste it when you have a severe cold and you will perceive hardly any flavour and hardly any smell. A coffee bean, on the other hand, is one of the most delicious smells, and savagely bitter to taste. These are extreme cases that tell us that flavour, smell and taste are bundled together in our heads in puzzling ways, and that the taste and the smell have meanings that sometimes chime together and sometimes ring quite different bells.

I have another puzzle, set for me by my younger son who was born blind. He has a really unusual mind, which combines a deeply intelligent understanding of classical music with an inability to understand many straightforward things like numbers, relationships and consequences. He is a 'savant': someone with severe learning difficulties and an outstanding talent. He cannot confidently add two and two, but ask him to improvise a new Chopin nocturne or a Bach two-part invention and he will deliver a brilliant piece of music without a moment's hesitation. His question about smell is 'What does three o'clock in the morning smell of?' I do not know the answer. Perhaps he is awake and perceiving real smells at three o'clock in the morning; or perhaps his dreams are olfactory dreams, while mine hardly ever convey a smell. In any case I would like to give him an answer.

Out of this string of little puzzles there grew in my mind a conviction that the sense of smell speaks a language that I would like to understand. It offers meanings at several different levels. It speaks of our immediate surroundings and tells us whether they are familiar or unfamiliar. It speaks of our personal past, of comfortable and disturbing experiences we have had. It speaks of our likes and dislikes. It offers pleasures that no other sense can offer. It speaks of home and abroad. It speaks of the inner nature of things and their affinities with other things.

What is the smell of fresh rain on warm earth and vegetation? What are the words to describe it? How could one convey the smell of summer rain to someone who has not experienced it? What are the words that would distinguish this smell from other damp and earthy scents like a boiled potato or a cellar or a damp tea towel? And why should the smell of fresh rain be so evocative, more powerful in reviving a complete sensation of the past than our other senses? And why should the smell of a summer shower have connections to the smell of mushrooms and beetroot and beaches and thunderstorms and cucumber?

You might have thought that someone somewhere would have dug through these questions and come up with some answers. This is, after all, a fairly obvious sense. It accompanies every waking breath we take, and quite probably many of our breaths during sleep. It is found in the most prominent place at the centre of our faces. Although less important to us and less charged with meaning than our senses of sight and sound, it is not so far behind that it should have been neglected. Most people, I guess, would give it a higher place in their sensory array than the remaining four senses of taste, touch, proprioception (the sense that tells us how our bodies are positioned) and balance. Some might argue that taste has a higher place, but I will demonstrate that smell beats taste in our perception of flavour by a long nose.

A search through the literature of smell shows that two aspects of this sense have been well explored. There are many books about perfume: some of them interesting, most of them passionate. Then there is a wealth of scientific articles about the sense of smell – its chemistry, its genetics, its physiology and psychology – but few books for the non-specialist reader. It is a pity this should be so, because there have been major advances in the science of smell in the last few decades, which deserve to be better known. It is only two decades since Linda Buck and Richard Axel won the Nobel Prize for unravelling the genetic code for the nerves that detect smell. On the back of their work, there is now a much better understanding, although not a complete one, of how our brains interpret information about smells, about defects in the sense of smell, about the role of smell in flavour, and much more. For someone like me, who wants simply to understand the meanings of this sense in my workaday experience of life, the literature is too specialist, too much about the perfume laboratory and too little about the everyday.

To be fair, there are a few topics other than perfume that have been dug over. Wine buffs have elaborated a Wine Aroma Wheel to describe all the combinations of aroma that a glass of wine may give off. There are 90 descriptors on the wine wheel, and that is a good sign that smells can be unpicked and described in fine detail. Whisky, beer and cigars have their aroma wheels too. The International Coffee Organisation has created an official list of descriptors for the aromas of coffee beans. And what else? It is astonishing that many things that give off delightful scents in interesting varieties have not been described. I can find no authoritative study of the scents of roses, though many people would immediately think of roses as a source of varied olfactory pleasures. Tobacco, whose primary function is to create pleasure through smell, is considered by its aficionados as a matter of taste and mouth-feel, not scent. Wood, which offers many delights beyond the well-known sandalwood and cedar, has not been considered for its varied smells. And so on.

The expedition I decided to make should, I felt, go into all the territory that the perfume and wine experts have left untouched. It should take my nose into the street, the kitchen, the seaside, the hedgerow, the garage, the city, the church, the farm, the country. It should take my nose to the scents of all the seasons. It should explore the territories of fruits and animals and grains and marine smells and earthy and metallic smells. It should allow me to savour and understand the meaning of such innocent delights as the scent of hot tar, of cinnamon, of lubricating oil, of ripe melons, of brown sugar, of bonfires, of lavender, of old garden sheds, of hardware shops, of beaches, of frying bacon, of gin, ginger, jasmine, juniper and jute.

As I made my preparations for exploring the sense of smell, I unearthed some prejudices which may have inhibited others from making the journey. One prejudice is that this sense is too primitive for words. Smell, according to this prejudice, is the main sense of inferior creatures – hagfish scavenging in the depths of the ocean, snakes in their holes, minnows in streams. Several philosophers have placed it amongst the lower senses, inferior to intellectual senses such as sight and hearing. The 18th-century German philosopher Kant for instance said:

> To which organic sense do we owe the least and which seems to be most dispensable? The sense of smell. It does not pay us to cultivate it or refine it in order to gain enjoyment; this sense can pick up more objects of aversion than of pleasure (especially in crowded places) and, besides, the pleasure coming from the sense of smell cannot be other than fleeting and transitory.

No, Herr Kant, I cannot agree with you. Smell is a sophisticated sense that deploys a very large array of specialised brain cells. It is the sense which discriminates most sharply between the known and the unknown. It communicates more directly with the brain than any other sense. It is the source of the most vivid memories. It is the sense which gives flavour to our food. The pleasure it gives, especially now that we live in a hygienic, well-drained world, far exceeds the aversion. Above all, it is the sense that puts us at ease with our environment. It is the sense that tells us that we are at home or abroad. For many people, the moment when they fully appreciate that they have left their home country is when they smell foreign cigarette smoke, foreign disinfectants, foreign cooking, foreign metros. The moment when they are sure of being on safe ground is when they smell their home. It is well known that Kant never moved from his cold home town on the Baltic, and so never exposed himself to such pleasures. Probably he suffered from a permanent sniffle, living in that cold, damp climate, and perhaps his sense of smell was impaired.

There is another prejudice that inhibits people from exploring their sense of smell. This is the belief that animals are so much better at smelling than humans, and that we cannot hope to be in their league. Indeed, I suspect there is a belief that the sense of smell is atrophying in humans: that our ancestors probably sniffed like dogs when they first took to the savannahs; that they lost the need to smell when they stood up and lifted their noses from the ground; that modern humans are left with a feeble relic, as unimportant to our lives as our little toes. Darwin took this view.[1] Freud went further and asserted that the more civilised a society, the less attention it paid to smell. But it is known that he had grave problems with his nose.[2] He anaesthetised it by sniffing cocaine and underwent several bouts of surgery, so perhaps he, like Kant, started with a personal prejudice.

It is true that many creatures have olfactory senses that perform tasks humans cannot perform, but that does not mean that the human sense of smell is feeble (and there are plenty of animals that cannot track and instead hunt the way humans do, by sight rather than scent). We have olfactory powers that perform tasks many creatures could not manage. Bloodhounds can pick up scent trails that are days old, and can tell in which direction the trail was laid; but there are probably no dogs that can discriminate, as humans can, between the scents of Grenache as against Syrah grapes in wine, or between the odours of petrol and jet fuel. Carrion eaters like vultures can scent minute concentrations of the smell of putrefaction at great distances. (Gas engineers in the USA have been harassed by turkey vultures mistaking the smell of leaking gas for carrion.) But I doubt that vultures could distinguish Earl Grey tea from PG Tips from tea made with milk that is slightly off, all of which would pose no problem to a human being. Mice can tell by smell which female is on heat. But I doubt they can distinguish the smell of lager beer from bitter, or Balkan from Virginia tobacco. The plain fact is that different creatures have evolved olfactory senses with very varied functions, and that humans have an excellent sense of smell for the purposes that matter to us. It is a myth that our sense of smell is defective because we cannot track by scent, or that it is redundant since we took to standing upright, or that it is third rate because our senses of sight and hearing are dominant.

There is indeed one thing that we humans do, with our modest noses and our sense of olfactory inadequacy, that other creatures do not do. It is called 'retronasal olfaction' (something I will explore in more detail later), and we do it many times a day without realising we are doing it. It is possible that we would be justified in thinking ourselves superior in our sense of smell merely for this one talent.

There may be another factor at work – prudishness. The fear of dodgy smells was a major worry for Kant, and may be a worry for others today. Americans, usually so bluff and direct, are cautious about smells. In the USA the word 'smell' is most often used as a disturbing verb rather than a neutral noun – some people smell, others use deodorant. Even the word 'odor' signifies something to be controlled – one's home should be 'odor-free' and one should buy products that 'neutralise odors'. The acceptable words are 'aroma', 'scent' and 'fragrance', words that seem coy to the English.

Mouth odour; body odour; foot odour; the risk of breathing in when in a crowded lift; the risk of eating beans; these things provoke social unease, or, the humorous adjunct to social unease, a snigger. Worse, they encourage writers on perfume to display their down-to-earth qualities by bluff accounts of psychological experiments to test whether women or men are better at identifying smells from old underwear (women are better), or whether mothers can identify their babies from their nappies (they can), or whether armpit swabs are attractive to the opposite sex (sometimes, although not in expected ways). Such experiments, usually conducted in university psychology departments and using first-year students as the subjects, are of doubtful validity, and seem to pander to the prejudice that smells are dirty.

For some people, there is another inhibition to thinking about smells. This inhibition stems from the very success of perfume and wine buffs at describing their particular subjects. The language of perfume marketing and wine lists is, for some people, language they could not possibly use themselves. They assume that this is the language to be applied to all smells, however down-to-earth. 'Floral heart notes are combined into voluptuous chords that are sultry, sophisticated, radiant, narcotic, exotic'; 'ambery orientals have a spicy top with a vanilla base'; 'a Chypre based on the contrast between bergamot and oakmoss with generous top notes of citrus'; 'heavy-earthy with a rich undertone of precious wood notes'.

This is not language that we could apply to freshly laid tar or fried bacon or seaweed or new-baked bread. We need language that is much less exotic and precious, that is more accessible and down-to-earth, that requires no special knowledge of obscure products like oakmoss or ambergris. You may think it a forlorn hope that we can find such a language. After so many millennia of human speech, we should surely have found it by now. It is my contention, though, that we already have the language to describe smells, and that it requires just a little extra concentration and a little technique to put smells into words. What we can describe has meaning; what has meaning gives pleasure.

It is a surprise to find that some people have succeeded in coming to grips with the sense of smell, and that they accord it as much importance and as many words as their other senses. There are for instance tribes living in equatorial rainforests who give the sense of smell the status that we give to our sense of sight. For them, living in dense vegetation with no long views, smell is the sense that best gives them information about things at some distance – ripening fruit, flowering trees, animals high in the canopy which may be invisible but are detectable by scent. Some of these tribes in the Amazon, the Andaman Islands, or Papua New Guinea weave the sense of smell into their entire understanding of the world.[3] Some tribes define life as the condition of having active smell, and death as the condition of having no smell. They also describe the changing seasons in terms of smell, the geography of the forest, the alienness of other tribes. Even in our more visual world there are people who have no difficulty describing smells. That is obviously true of the professions whose products have valuable aromas – distillers, perfumers, brewers and coffee blenders. It is also true of a few great writers who have recognised that smells carry a powerful emotional charge and who take the trouble to describe them in simple words as an essential part of bringing their imagined worlds to life. Such writers are few, it is true, and they have made a particular virtue of putting sensations into words. They are vastly outnumbered, however, by the writers who dodge the problem of describing smells, and shelter behind the indefinite – 'that unmistakable, indescribable smell'.

As to the smell of fresh rain itself, there is a scientific account. In 1964 a pair of Australian scientists called Bear and Thomas published an article in *Nature* called 'The Nature of Argillaceous Odor'.[4] In it they coined the word 'petrichor' to describe the smell of fresh rain: from the Greek *petra* (stone) and *ichor* (the blood of the gods). They identified that one of the main components of this distinctive smell is an oil trapped in rocks and soil, which is released when it rains. This oil is secreted by some plants during dry periods, and serves to halt the growth of roots and to prevent seed from germinating. In moist areas another ingredient in the smell of fresh rain on earth is geosmin: a chemical produced by soil-dwelling bacteria known as actinomycetes.

So the smell of fresh rain is in fact a blend of tiny, invisible, airborne chemicals given off by plants and bacteria in the earth. As we will see, most smells in nature are a blend of diverse components, and it is difficult to sort them out. As we will see too, some of the components are given off by quite different sources, and they create unexpected associations in our heads. Geosmin, for instance, is a component in the smell of fresh rain, but we can also detect it in the earthy smell of beets and spoiled wine.

This account seems like the end of the story. Putting a chemical word to a smell has an air of finality – no more need be said. But it is not the answer; it is just a label for a chemical. We do not talk in chemicals, we talk in plain words, and in describing smells we make use of comparisons to things we have encountered in plain words. We want to know what a smell tells us about our environment, about our personal memories, about the associations that it evokes, about what the smell is 'like', more than we want to know about its chemical composition. On top of that, the scientists' account is not even a complete description of the mix of chemicals that make a smell. When the composition of a scent is analysed, it is typically found that dozens or even hundreds of different types of chemical are being sniffed, and that quite an insignificant subtraction or addition of components will alter the character of the smell. So we recognise smells not as a mix of separately-identified components, but as a 'chord' that makes an odour we can recognise, a smell with a meaning. Not petrichor and geosmin, but the smell of fresh rain. James Joyce, one of the writers who understood the sense of smell, has this description of rain on earth which I find poetic: 'the rain-sodden earth gave forth its moral odour, a faint incense rising upward through the mould from many hearts'.

Note that there is a touch of stagecraft in the naming of some chemicals. Geosmin is simply a Greek label for the smell (*osme*) of the earth (*geo*). We find the same trick again and again: for instance with cadaverine or asparagusine or putrescine. These names mean nothing more than 'the chemical that smells of the smell in question'.

My expedition of discovery into the sense of smell would, as I saw it, not be a heavyweight scientific expedition, burdened with measuring equipment. Rather it should be a personal, lightweight expedition – just my nose and a sense of curiosity about all the different aspects of the natural and man-made world which offer interesting scents. I would as far as possible take my nose off the beaten track of perfumes and wines, and head into unmapped territory. I hoped to return with a rough map of the territory and a rough way of describing the many smells I would encounter. It was not likely that mine would be a map that was complete and satisfactory for everyone. Our sense of smell is too personal for that, our olfactory blind spots are too numerous. Each of us is sensitive to different smells to varying degrees. The associations conjured up for each of us by smells are personal and difficult to share with others. But at least my expedition might encourage others to tackle the puzzles of smell more thoroughly.

First, in the spirit of all expeditions, I had to get my nose in training. I decided to start in a place whose scents and associations and whose very air would be fresh to me.

CHAPTER 2

The Salty Tang of a Sea Breeze

'Sea Breeze – Always mysterious, vast and deep, the sea has been celebrated in song and story through the ages. This fresh, crisp, exhilarating scent captures the clean salty tang of a brisk sea breeze'.

Internet advertising for fragrant oil

I decided to start my exploration at a sea port. In search of first-hand experience of seaside smells, I took the train from London to Portsmouth Harbour. For those like me who love the sea and ships, Portsmouth Harbour is an exciting destination. It combines the solemnity of ocean-going ships with the tacky cheeriness of seaside places and the faded glory of the Royal Navy, which has used the harbour as one of its main bases since Tudor times. Being beside the seaside would give me chance to clear my head of the habitual smells of my London home with some fresh sea air. It would test my ability to grasp the fleeting scents of a working port and it would give me a first chance to interpret the meanings of smell.

The station platforms at Portsmouth Harbour are built out over the water. Between the rails you can see the green heave of water and tatters of seaweed hanging from the station's metalwork. You expect the first smell as you step down from the train to be the sea – a salty tang. And the first smell is indeed the sea – but it is not salty and not tangy in the expected way. The smell is mineral and muddy and seaweedy. The dominant smell is chalky, no doubt from the shells of small crustaceans and molluscs, and from the chalk of the coastal hills which has made the water an almost Venetian pale green. With the chalk, there is a sour, rich mud smell with something cabbagey and oily about it. It is an English Channel smell, and quite different to the scent of the Atlantic and different again to the Mediterranean.

Standing on the platform, obstructing the other passengers hurrying to the exit as I fumble for a description of the sea's scent, I am brought face to face with one of the first difficulties of olfaction. It is no use sniffing on and on, in a desperate hope that the smell will reveal itself if I concentrate hard enough and sniff often enough. Quite the reverse: repeated sniffs rapidly dull my appreciation; after three or four sniffs I have become habituated, and I have ceased to register the smell. 'A single natural sniff provides as much information about the presence and intensity of an odour as do seven or more sniffs', wrote the Australian psychologist David Laing,[5] who conducted the definitive research on sniffing. Sniffing longer and harder does no good either. The first fraction of the sniff does the trick, and a longer sniff just confuses the mind. In fact, scientists have measured the time it takes to register an odour: the first 160 milliseconds are all it takes until the brain has perceived the smell and is reacting to it. That is a sign of how fast the neurones – the electrical wiring of the brain – take to transmit a sensation, and how little is added by prolonging the process.

So, although it seems that more sniffing is better, that more effort, and greater concentration will be repaid with better perception, the opposite is true. Short exposure, followed by a change of smell, allows one to appreciate the original smell better. That is why people in the perfume industry refresh their noses between smells. Perfumers have variously used ammonia or camphor or coffee beans to refresh the nose during long smelling sessions. Coffee beans are now standard issue in perfume retailing, and are part of the stagecraft of selling a luxury product. In ordinary life, we do not have these accessories to hand, and must find simpler ways of refreshing our noses.

All the passengers had by now left the platform at Portsmouth Harbour, and I was still standing puzzled, searching for the salty tang of the sea. Certainly the air has a tang. But it seems more the *feel* of sea air than a smell. It is the feel of chill and freshness from a September breeze blowing across the harbour. It is not the smell of salt, except in the imagination. In researching smells, I have been forced to conclude that refined salt has no intrinsic smell; it is a taste on the tongue, and if we think that we smell it, that is usually by association or because a small dusting of the salt has been breathed through our noses onto the taste buds on our tongues. No, the smell of the sea is not salt, nor yet ozone, it is mostly dimethyl sulphide, a by-product of plankton, and its smell has a touch of the offensive cabbaginess of sulphur.

Out at sea, a sailor will not notice the smell of the sea: where the dimethyl sulphide gas is emitted in mid-ocean, it rises into the marine clouds and is distributed around the planet. He will only notice it on approach to land, where it is caused by the reaction of dimethyl sulphide with seaweed. The tang of the sea should more properly be called the tang of the shore.

Out of the station, past the sweet, light smoke of cigarettes, past a thin thin man whose thin thin cigarette smells sweeter still and more grassy, down to the jetty for the ferry, to cross the harbour from Portsmouth to its sister town of Gosport. The ferry boat looms across the harbour, a stubby, low vessel with rounded bows and stern, like a nautical dodgem car. It sidesteps an ugly little tanker – *Jaynee II* – trundling out of the harbour with a smear of fuel oil exhaust behind it. Out of a neighbouring quay a seven-storey ferry manoeuvres with a rumble of mighty marine engines echoing in lumpy metal. As it turns to clear the harbour entrance, it shows its strange shape – it faces both ways being symmetrical fore and aft – and it shows its name: '*St Clare*'.

The Gosport ferry arrives, touches the floating jetty with a shock, followed by an aftershock as the jetty rebounds off its uprights. I step on board over the rubber lip of the ferry boat. Onboard there are some of my favourite smells. There is a smell of varnish from the wooden benches around the inside – resinous, warm and sticky. There is an engine-room smell of rich, deep fuel oil like a mineral molasses with a sulphurous edge, a scent of lubricating oil, sweeter and lighter, dripping from every vibrating joint, and the permanent stuffiness of hot air. Up on deck, where I position myself near the stubby square funnel, there is the utterly moving smell of marine fuel exhaust – rich and homely, a blend of stewed fruit and seaweed and hot metal – alternating with the sour freshness of seaside air. It smells like every sea voyage I have ever made.

This sense of familiarity and nostalgia is one of the most telling effects of the sense of smell. It connects me to my other sea voyages, long and short, portentous and trivial, emotional or matter-of-fact, across the Mediterranean or across a Cornish estuary. Familiarity and unfamiliarity are sensations that so often accompany the act of olfaction that I am driven to conclude that they express one of the central functions of the sense of smell for human beings. More than any other sense, smell has a special power to evoke memories. It owes this power to its location inside our skulls: close to the emotional centres of the brain and its structures for laying down long-term memories. More than any other, smell is the sense that puts us at ease with our environment, that tells us what is familiar or unfamiliar. It is the sense that reassures us that we are on safe ground and eating safe food, or, conversely, in the presence of the unknown and dangerous.

This trip across Portsmouth Harbour is one of the shortest maritime voyages of my life, perhaps 700 metres of chalky green water from Portsmouth to Gosport. But it is long enough to evoke the solemn emotions of a real sea voyage. I am moved by the moment of departure from the jetty when it seems we are in motion and not in motion, when the first inch of separation becomes a foot and then a yard of swirling water. I am moved by the sensation when, clear of the jetty, one looks over clear water and feels the solemn presence of the sea, by the judder of the little ship's metalwork, by the lumpy paint and flakes of rust, by the sight of a seagull at home on the unhomely sea. Perhaps most moving of all the sensations, though, are the smells coming in gusts from the funnel and the wind.

The voyage is over in a trice, and I step off the boat in Gosport in search of a boatyard. I have a particular boatyard smell in mind – sweet sawdust, rich tar, grassy rope, intrusive paint. The reality is quite different. Modern chemistry has replaced the old smell of the boatyard, which would have been familiar to a Victorian, or a Tudor and even a Viking boatbuilder alike, with new smells. Fibreglass, nylon rope and modern paints are not a patch on the old materials for smell.

I walk along the perimeter fence of Gosport boatyard with very little reward, just the occasional sickly-sweet smell of buddleia growing in untended corners, and the whine of an electric sander. It is difficult, this smelling business. In the normal run of life we give a low level of attention to our noses. We are constantly alert to what we see and hear, and find it difficult to switch those senses off. But our sense of smell is rarely at the forefront of our minds; more usually it is just below our threshold of attention and it takes a conscious effort to register what we smell. We give most of our attention to sights and sounds, which convey meanings that are immediately important to us, while those other senses, like smell, which convey less accessible meanings, are given less attention. Indeed, as we shall see, if the sense of smell were to be put in the centre of our attention, it might get in the way of the kind of thinking to which our brains are so well adapted. Furthermore, any smell that attracts our attention is fleeting: we have no sooner breathed it in than we breathe it out again, and the next breath is accompanied by a different smell. In the gusty air of the seaside, the wind brings a hint of seaweed and then snatches it away; a flicker of wood shavings, and then it has gone; a half-breath of sour mud and then nothing.

Then I get a scent that shivers my spine. I seem to recognise it, but I cannot place it. It seems to have come from somewhere in the memory, but I cannot associate it with any particular episode. This is the tip-of-the-nose effect which we experience so often, of a smell that we almost recognise, but we cannot quite pin down, a smell we can almost describe but the details of which seem to elude analysis.[6] The smell is aggressive, bitter-sweet, akin to the smell of gas and of glue. I guess it must be epoxy resin, used to patch the fibreglass hulls of boats.

At a loss, I stand wondering what to do next and twirling a strand of dry seaweed in my fingers on the edge of the boatyard. Suddenly there is a bloke in oily overalls, who wants to know what I am doing. Does he think I am a spy, studying equipment on Royal Navy ships; or a suicide contemplating my last walk into the water? He is certainly puzzled by my explanation that I am investigating the sense of smell. 'What do you want to come to Gosport for, then? I'd have thought there would be better places for smelling.' I try to explain about the smells of the seaside and boatyards, and the delights of rope, tar and paint. He introduces me to his two mates, also in oily overalls, and all three of them give me the indulgent look of people dealing with a friendly sniffer dog. These are boatyard workers with a chunky mobile crane for lifting boats out of the water, and together we have a good sniff at it. The three of them take to the business of sniffing their crane with good humour. 'I'm getting the heady fragrance of diesel, vintage 1999.' 'Mmm, a fruity, raspberry note.' 'That's not the diesel, mate, it's your overalls.' 'A trifle young, perhaps, this lubricating oil.' 'A cheeky young Esso.' They have caught the effeteness of wine lists, and are enjoying applying it to their dirty old crane. Lubricating oil, synthetic rubber, diesel, well-known smells. But also a new smell: it is hydraulic fluid. I am allowed to sniff a ten-litre can of the stuff – it is a sweetish, light mineral oil with a pronounced touch of apricot and fig. I feel more than a little like a dog, lowering my nose into a can and sniffing, watched by three smiling men.

Back I go towards the ferry terminal and the seafront shops. I pass a succession of fast-food cafes, each with a couple of electric wheelchairs outside, and each with a ventilator announcing to the street its particular aroma of used cooking oil.

Were I Sherlock Holmes I would write a treatise on a hundred different odours of cooking oil, as Sherlock wrote a treatise on a hundred varieties of cigarette ash. Coriander-flavoured smells announce a take-away curry shop; a vinegar flavour announces a fish and chip shop; and so on. I decide to experience the smell (but not the food) of McDonald's. Like most complex smells, this one comes in many layers, and each one can be unpicked. The dominant odour here is vinegar: bitter and invasive. Under that is burnt vegetable oil, whose cloying syrupy warmth is halfway between mineral oil and compost. Under that is potato: not crispy fried potato but a soggy, muddy, beetroot smell. In occasional wafts I also smell strawberry, Brylcreem (many Gosport men have old-fashioned slicked hair) and turpentiney, powdery perfume. The people in this McDonald's have a look that I associate with broken nights – heavy upper lids to their eyes and a dark line below the eyes – but maybe it comes from the diet.

I am beginning to learn how to distinguish the smells of the seaside and what words fit those distinctions. I had launched my expedition uncertain whether I would be able to make any sense of smells, and even less sure whether they could be put into words. Now I am sure that it is worth the effort: there are meanings in smells that can be brought to the surface, and there are differences and resemblances that can be put into words. Some odours are complex, with several layers; others are simpler. Tar and marine mud and impregnated wood and natural rope are complex and layered; vinegar and cigarette smoke are simpler. Some smells are rich; some are thin. Fuel oil is rich; lubricating oil is thinner. Some are sweet, some are bitter, and many are bitter-sweet or sweet with an edge. Our noses make much more subtle distinctions between the varieties of bitter and sweet than do the taste buds on our tongues. Cannabis smoke is sweet; epoxy resin is bitter; strawberry flavouring is sweet with a sickly edge; fresh paint is bitter-sweet. Some smells are fresh; some are old. Newly sanded wood is fresh; a baulk of wood that is supporting the keel of a boat in a boatyard, and has probably supported the keels of many generations of boat, smells old. We can also distinguish a few aggressive smells that get up our noses – piquant smells like vinegar, like epoxy resin, like the higher fractions of petroleum. The common feature of piquant smells like these is that they excite the specific nerve in the face – the trigeminal nerve – that also responds to pain and cold as well as piquancy, and that makes our eyes water or our noses sneeze. These distinctions offer the foundations of a way of describing smells, a vocabulary of odour that I am sure can be developed to describe the full range of smells.

On the return journey in the ferry across the harbour to Portsmouth, I can revisit the sensations of the ferryboat. Once again I stand on the little deck taking pleasure in the marine engine scents and the sights and sounds of a busy harbour. It seems to me that there is a special bonus that comes from concentrating on the hidden sense of smell. The bonus is that the more insistent senses of sight and sound continue to make themselves heard as before, but they are enriched. Being attentive to smells makes me more attentive to all my senses, and able to savour the whole experience of a boat journey in all its dimensions. The senses amplify each other, and bringing the sense of smell into play amplifies the other senses.

There is another bonus. On my return to Portsmouth, I wander into the Royal Navy dockyard, now a heritage site. Here at last I find the smell that I came for. In an old lofty workshop there is a repair yard for old wooden boats. One step inside and the air is full of the fragrances I want. It is a rich cake of smells: smokey tar, fermented meat, sticky varnish, dried fruit, paint, hemp – all blended from the wood and rope and half-used tins of stickiness that lie around. The Vikings would have felt at home.

This has been a short reconnaissance into the sense of smell, a first trial exploration. What have I learnt? First I have found some important truths about the mechanics of smelling in the real world, where sensations come at us fast and fleetingly, and where all our senses are competing for attention. It is difficult to grasp a smell. It comes and goes too rapidly for analysis. It hovers on the edge of recognition and then takes flight. It comes out of thin air and goes into thin air. Wind blows it away. Cold freezes it.

Our minds too put obstacles in the way of appreciating what our noses are sensing. Our minds look for meaning, and find plenty of meaning in what we see and hear. So smell, which certainly has meanings but less obvious ones, tends to take third place to vision and hearing. When a smell gets through to our brains, we rush to recognise it, so fast that we cannot dwell on it and pick it apart. But we shall see that the meanings of smells can be teased out, and can enhance our experience of life. I have also learned that it is possible to bring the sense of smell higher up my consciousness, to give it attention without cutting off my more insistent senses. The right frame of mind for making sense of smells is one of relaxation, of receptiveness, of letting the present moment announce itself. It is the opposite of straining after meaning. It is a frame of mind, of mindfulness, that allows me to be more in the present, and it gives an extra dimension to the pleasure that all the senses yield.

This seaside trip has shown me that the problem of the neglected sense of smell is beginning to yield to exploration. There are meanings in smells that can be understood by concentrating on them. There are ways of overcoming our normal, hurried lack of attention to smells. There are words for smells. All this tells me a little about what my brain does with a smell once it is perceived, but I need to go deeper: I still do not know how a smell gets there in the first place. I need to know what a smell is, whether it is simple or complex, orderly or disorderly, how our noses process it and how our brains recognise it. How do our sensory nerves interpret a smell and turn it into a perception? And is there a regular pattern to smells, so that we can locate one odour alongside another, like colours in a spectrum or notes in a musical octave?

CHAPTER 3

Metals and a Cup of Coffee

'"Here you are, doggy! Good old Toby! Smell it, Toby, smell it!" He pushed the creosote handkerchief under the dog's nose, while the creature stood with its fluffy legs separated, and with a most comical cock to its head, like a connoisseur sniffing the bouquet of a famous vintage.'

Arthur Conan Doyle – *The Sign of the Four*

In search of the inner nature of a smell, I found myself following a curious trail along the smells of metals. Blood smells metallic. So do onions, electric trains, sweat, garlic and blue cheese. What then is the common smell of metals, and what is the difference between copper, steel, aluminium or zinc? Straight away we encounter a curiosity. Everyone is sure that there are metallic smells, but the strange fact is that metals do not, on analysis, have an intrinsic scent.

I visited a local trader in small metal things – secondhand tools, keys, locks and bolts. There were metal fittings, lovingly polished, hanging like grapes around his shop. 'Do the different metals smell different from each other – steel, copper, lead, brass?' I wanted to know. 'Sure, they smell different.' I wanted to know how each metal smelled to him. 'Well, if you scrape it, steel smells different.' 'And copper?' 'Sure, copper smells coppery, quite different.' 'And lead?' 'Yes, it smells different, like a lead smell.' 'And brass, and zinc?' 'Brass smells very brassy, and zinc, well that's different too.' I put my nose to some of his secondhand tools: all I got was the smell of light oil.

Little the wiser, I dropped into our local shop where garden tools are renovated and sharpened. I brought with me from home a bundle of shears and scissors for sharpening. It was a delightful old shop, encrusted with oil and metal filings, with a rank of lawn mowers outside ready for collection. Secateurs, shears, knives and scissors, chains and random spare parts hung from hooks on every wall. The owner had taken over the shop from his father, having occasionally worked there alongside him as a boy. Yes, he told me, he was sensitive to the smell of metals, especially when filing old metal. 'A fine old pair of shears, when you file it and get that smell of old steel, it smells of bacon and onions. It takes me straight back to when I worked alongside my dad.' For him this was a Proustian moment, when a smell resurrects a long-hidden memory and brings it back to life, more powerfully than the sight of a pair of shears or the sound of the metal file. It was the effect Proust captured when he wrote of the smell of a little French sponge cake, a madeleine, dipped in herb tea, and recorded the sharpness of olfactory memory. Psychologists call it 'a madeleine moment'. The connections between smell and memory are more intimate than for our other senses, and I will return to them at greater length.

I related the science of metal smells to the man sharpening shears: how when scientists analysed the gases given off by metals, they found that the molecules in the gases that were responsible for the smell were not molecules of the metal itself, but were the molecules produced by human sweat interacting with metal.[7] We are smelling not metal, but the effect of our own sweat as it corrodes metal in minute quantities. To prove this, take a packet of metal things (nails, screws, staples) that have not been handled. Open the packet and sniff it, then take a few of the objects and warm them in your closed fist. The difference will be obvious. To take one metal – iron – as an example: when the gas given off by iron in such a contact is analysed at a molecular level, it is found that much of the smell is being made by one of the organic molecules present. This molecule is called 1-octen-3-one, and has a recognisably 'metallic' smell with a fungal tinge to it. The same molecule is given off when blood comes into contact with skin, since blood contains iron, and the contact with the sweat on our skin produces a similar chemical reaction with the iron in blood, and a similar metallic smell. The same molecule is given off by garlic, and mushrooms and blue cheese. Other metals give off smells in the same way. Copper and brass (the alloy of copper and zinc), produce organic molecules (not metallic) from contact with skin, and these also have smells which we identify as metallic. Zinc is more hesitant to form a metallic odour, and has to be rubbed very strongly against skin to produce the same effect. But in all these cases, it is not the metal evaporating, but the decomposition of skin acids by metallic molecules, that produces the molecules that, when we smell them, we recognise as metallic. The scissors and shears man did not believe it. Why should old steel smell different to new steel? Why should our perception be of a metallic smell, not of a sweaty smell?

However, the scientific analysis of metallic smells is as good a point of entry as any into the way this sense works. It reveals that smells are made by molecules floating in the gas given off by substances that are then detected by nerves in our noses. Not just any molecules, but the light, organic molecules which are the molecules of life. Molecules are the tiny clusters of atoms bonded together that make the smallest fundamental unit of a chemical compound. They can be very simple and very complex; very smelly or entirely odourless. A molecule as simple as ammonia (one nitrogen atom bonded with three hydrogen atoms) packs a powerful punch in your nose. A molecule as complex as a strand of your DNA will not give a smell at all because it is too heavy to float off in a gas. We detect the molecules of smell with nerves inside the very tops of our noses, situated near our eyebrows. Molecules are for the most part so tiny that they cannot be seen even with the most powerful conventional microscope: they are smaller even than the wavelength of light, and conventional microscopes depend on light. Yet, small as they are, our noses can detect them. They are breathed in, they stay only as long as we can hold our breath, and then most of them are breathed out again. The next breath brings a new cluster of molecules and a new smell.

Which molecules make a smell? To start answering this question, we would do well to pick apart the smell of something common – a cup of coffee will do. The tool for picking apart gases is called a gas chromatograph, invented in the 1940s. It is essentially a device for getting molecules to race round a track, with built-in handicaps to make the heavier molecules move slower. By the end of the race, the molecules are strung out so that you can identify each type as it passes the finishing post. A gas chromatograph is a very long, thin tube coiled up, with a heater under it to make the molecules volatile. A stream of helium gas drives the active molecules along the tube. The inside of the tube is coated with a gel that holds the molecules back as they progress along the tube. To analyse a gas, the temperature in the gas chromatograph is progressively increased, making each kind of molecule volatile in turn depending on their weight. Molecules of the same kind work their way along the tube at much the same rate, and the slower molecules are held back more by the gel coating. The result is a spaced-out succession of groups of molecules emerging at the end of the tube, where they are recorded. They can then be sniffed to tell us what they smell of, and whether they make a perceptible contribution to the smell. Chemists have developed a simple language for describing the smell of individual molecules – 'buttery', 'putrid', 'clove-like', and so on – but as we shall see, the smell of a fragrant substance is much more than the sum of its component molecules. At the start of gas chromatography, people would do the sniffing, but nowadays a computer can predict what the molecules are, and so what they smell of.

Let us take a cup of real coffee, brewed from roasted Arabica for instance, capture the smell it gives off ('headspace capture' in the language of the analysts), and run it through a gas chromatograph. The molecules that emerge from the device are very varied indeed. At least 800 different types of molecule are given off by a cup of coffee. They can be identified using another device – the mass spectrometer – which sorts molecules according to the ratio of their mass to their electrical charge, and so allows scientists to work out their chemical structure. Many molecules contribute nothing to the smell, and some are present in tiny quantities below the level at which we can perceive them. Nevertheless, when the odourless molecules are eliminated, we find that the smell of that cup of coffee is created by 27 different molecules. It is very much a multiple thing, not a single thing. There is not a single master molecule, a 'coffee smell' molecule, but instead there is a cluster of molecules whose pattern is recognised in our brains as the smell of coffee. That means we are doing two separate and subtle things to extract the meaning of a smell – we are both detecting the cluster of tiny molecules and we are recognising the pattern. The detecting happens in our noses; the recognising happens in our brains.

Only a chemist will derive much sense from a list of the 27 molecules that make up the smell of roasted Arabica coffee. But I list them here* to demonstrate three special features of

---

* Odorant molecules given off by roasted Arabica coffee: source B Bonnlander, Trieste Coffee Cluster (these descriptions have been standardised by scientists working in the flavouring and perfume fields)

• methanethiol	putrid, cabbage-like
• 2-methylpropanal	pungent, malty
• 2-methylbutanal	pungent, fermented
• 2,3-butanedione	buttery
• 2,3-pentanedione	buttery
• 3-methyl-2-buten-1-thiol	animal, skunky
• 2-methyl-3-furanthiol	roasted, meat-like
• mercaptopentanone	sweaty, catty
• 2,3,5-trimethylpyrazine	roasted, musty
• 2-furfurylthiol	roasted, coffee-like
• 2-isopropyl-3-methoxy-pyrazine	pea-like
• acetic acid	vinegar-like
• methional	cooked potato
• 2,ethyl-3,5-dimethyl-pyrazine	roasted, musty
• (E)-2-nonenal	fatty
• 2-vinyl-5-methyl-pyrazine	roasted, musty
• 3-mercapto-3-methyl-butylformate	catty, roasted coffee-like
• 2-isobutyl-3-methoxy-pyrazine	paprika-like
• 5-methyl-5H-6,7-dihydro-cyclapentapyrazine	peanut-like
• 2-phenylacetaldehyde	honey-like
• 3-mercapto-3-methylbutanol	soup-like
• 2/3-methylbutanoic acid	sweaty
• (E)-B-damascenone	honey-like, fruity
• guaiacol	phenolic, burnt
• 4-ethylguaiacol	clove-like
• 4-vinylguaiacol	clove-like
• vanilline	vanilla-like

the chemistry of smells. Firstly, these molecules are *all* organic molecules: the components of life, the molecules found in living things and usually built around atoms of carbon arranged in rings or long chains. They are for the most part made from a very limited palette of elements – of carbon (the great chain-maker of organic life), hydrogen, oxygen, nitrogen, sulphur and, occasionally, bromine and phosphorus. You might have expected to encounter some of the elements that smells call to mind – like a metal, or like chlorine, since we often recognise metallic or chlorine smells. But no, the huge variety of smells are constructed from the simplest ingredients, each molecule having particular properties, that are different even to those of molecules of near-identical construction.

The second feature of smell molecules is their variety. They are constructed, as noted above, from a small palette of atoms, but it takes only an insignificant shift in the number or order of the atoms in a molecule for it to change its smell. The reverse also happens – two molecules with very different structures can elicit the same smell. You could shift the order of atoms in a molecule, and you would have an entirely different smell, much as a minor change in the order of letters in a word will result in an entirely different meaning. And what the new molecule will smell of is largely unpredictable. Research chemists play around with molecules, most commonly in order to produce synthetic odours for the perfume or the food flavouring industry. They create new molecules by a little addition here, a re-ordering there, and they can never be sure what the next new molecule they create will smell of.[8] They may have hoped for violets, and instead the new molecule smells of grass or of nothing at all. The difficulty of predicting what a newly constructed molecule will smell of is well illustrated by the instance of two molecules with exactly the same constituents in exactly the same order, but one organised in a right-handed way, the other in a left-handed way. They smell different.

Two molecules can be mirror images of each other. In the same way that our two hands can have the same structure, the same order of bones and tissues, but the two thumbs stick out on opposite sides, so it is with some molecules. A molecule called carvone, for instance, comes in two versions, both with the same atoms and the same structure, but in one the oxygen atom sticks out to the right and in the other it sticks out to the left. The two versions of carvone smell different. One version smells of caraway seeds, and is a major component of the smell of dill, mandarin orange peel and the German liqueur *Kummel.* The other version smells of spearmint, and is a major component of the flavour of chewing gum. There are plenty of these mirror-image molecules; some smell different, some smell identical, some smell the same but with different intensity. It is astonishing that such minor differences between molecules should create such very different properties. It is astonishing too that our noses, ineffective as they may sometimes seem to us, should be able to discriminate between such minor differences in such tiny things as molecules.

The third feature of smell molecules is that the same molecule will often show up in quite different smells. Our cup of coffee, for instance, contains the molecule guaiacol, which also occurs in salami and cigars. Similarly, the molecule (E)-2-nonenal in our cup of coffee also occurs in cucumbers and the smell of old people's skin. These repetitions of smells are, to my mind, one of the delights of the world of olfaction. They give rise to the intriguing sensation that an odour reminds us of something quite different. Bourbon roses remind us of suntan lotion and coconuts, because they all share a common molecule; mushrooms remind us of oysters and earth, again because of a shared molecule. The world reveals an unexpected weave when we pay attention to the detail of its odours.

So how do we detect the molecules of smell? When we breathe in, we draw air up our noses, with some of the air passing through a narrow cleft into the very top of our nasal cavity. There, on each side of our nose, are the receptor nerves which respond to smelly molecules. Each patch of nerves is about the size of a small postage stamp, and is densely packed with about 6 million nerves altogether. Each side of the nose communicates with the brain separately, so you can experience 'perceptual rivalry' when each side of the nose is smelling something different. You could demonstrate the effect to yourself by putting competing scents under each nostril – for instance a coffee bean on one side and a clove of garlic on the other – and you would find it difficult to sort out which smell you were perceiving.

What has been described so far, however, is the orthodox route to smelling, but there is another route, which is used every time we eat or drink. We are, most of us, unaware of it happening. This other route is called retronasal smelling; it happens not on an in-breath, but when the action of chewing and swallowing pumps air into our noses from our mouths up the back areas of our noses. It is the source of flavour, and it is of huge importance in our lives, but we mistakenly assume that it is simply the sense of taste, and we mistakenly place the sensation in our mouths. It is a form of smelling that appears to be more developed in human beings than in other animals, and it will receive more attention later in this book.

Our millions of olfactory receptor nerves come in 352 varieties. Each variety of nerve is activated by one limited group of molecules only. They are fairly specific, but not so specific that they cannot respond when a new molecule comes along. Nor are they so specific that a molecule may not activate more than one type of receptor nerve. But the rough picture is that our in-breath brings a cluster of molecules up our noses, and that that cluster of molecules activates a particular pattern of receptor nerves. The brain then has to make sense of the pattern.

There is a heated and unfinished debate amongst scientists about how exactly a receptor nerve interacts with a molecule. The conventional theory was that the molecule slots into the receptor nerve like a key into a lock. That theory was dominant through most of the 20th century, but it has difficulty both in explaining how our noses can detect new molecules that they have never encountered before, and how they can tell the difference between the smells of two molecules with identical components distinguished only by the fact that one is 'right-handed' and the other 'left-handed'. A modified version of this shape theory suggests that different receptors detect only small features of molecules, and that these fragmentary inputs are combined to form a larger sensation of smell. Two less conventional theories come at the problem from quite different starting points. One theory contends that it is the volume of the molecules that matters, and that the receptor nerves are tuned to particular molecular volumes. Another theory, the vibration theory proposed by Luca Turin (in one of the few entertaining books on the chemistry of smells, *The Secret of Scent*), posits that odour receptors detect the frequencies of vibrations of molecules in the infrared range. Were that the mechanism by which our nerves detect smelly molecules, it would make the sense of smell more like the sense of hearing, which also detects vibrations, and it would make the comparison between smells and musical chords less of a metaphor and more of a real description. As yet, however, there is no generally accepted theory that explains olfactory perception completely.[9]

The route from nose to brain goes through a thin plate of the skull, just behind our noses, called the cribriform plate. For a very long time it was not certain where the actual smelling took place. Up until the 19th century, anatomists examining skulls looked at the thin cribriform plate, riddled with tiny holes, and thought that scent molecules themselves passed through the holes into the brain, and were smelled within it. In fact the perforations in the cribriform plate allow not the molecules, but the host of nerve fibres from the smell receptors in our noses to enter the brain. It is a short route, shorter than those taken by the nerves that serve our other senses. Each of the receptor cells in our nose is connected via a single synapse (the junction between nerve cells) directly to the brain.

The route from nose to brain is also a very fragile one. If you were tempted to thrust a biro up your nose with force (not to be advised) you might easily break through the cribriform plate and into the brain. More plausibly, a quite small shock to the head from a fall or a car crash can jolt the cribriform plate and in the process sever all the nerves that pass through it, leaving the victim with no sense of smell. This thin bone can also offer access through its many perforations to some dangerous viruses capable of attacking the brain. So, as well as allowing us to sense the molecules of smell, one of the most important functions of our noses is to filter and clean the air that we breathe. That explains why mucus is such a feature of the nose – it is our means of transporting dirt away from a sensitive organ and of disposing of that dirt through our guts. The mucus acts also as a solvent for odour molecules, flows constantly, and is replaced roughly every ten minutes. It may seem, when we have a cold, that mucus is a hindrance to smelling, but in normal conditions it is an essential lubricant of the system.

Their vulnerability to attack explains another special attribute of the nerves of smell. Most nerve tissue is difficult to replace, but the nerves for smell are an exception. The receptor nerves in our noses and the nerves in the brain with which they connect are constantly regenerating. They are in a constant state of regrowth, and we replace all the nerves of our olfactory system every three or four months. When surgeons want active nerve tissue that will grow to replace damaged nerves in other parts of the body, they turn to the nose. In a recent, ground-breaking operation to regrow the nerves in a man's severed spinal cord, surgeons used the nerves in one of his olfactory bulbs to create new nerve tissue in his spine.[10] Constant regeneration of the nerves of smell imposes some costs on us. For one thing, the nerves may not regrow exactly to plan, and we may as a result develop some blind spots (called anosmias by experts) or some false perceptions (called parosmias) which result in our brains misinterpreting a smell and confusing, say, bananas for burnt toast. For another thing, the sense of smell is liable to wear out in old age, because the nerves have regrown so often.

Already, before our exploration of the sense of smell has reached into our brains, two of the many meanings of smell have been revealed. Smells are made by the volatile molecules given off by a substance – be it the swirl of molecules given off by a cup of hot coffee or the few grudging molecules given off by metals that have been handled. They are bits of the thing itself, not the incidental effects on light or air pressure that create vision and hearing. A second meaning of a smell is that some of the molecules are shared with other substances, and allow us to perceive the intimate chemical connections between things – be it the connection between coffee and cucumber or between metal and blood.

There is much more to unearth about how the brain is organised to respond to smells. But first I wanted to sniff some fresh country air.

CHAPTER 4

Charcoal Burners

'Today I think
Only with scents – scents dead leaves yield,
And bracken, and wild carrot's seed,
And the square mustard field;

Odours that rise
When the spade wounds the root of tree,
Rose, currant, raspberry, or goutweed,
Rhubarb or celery;

The smoke's smell, too,
Flowing from where a bonfire burns
The dead, the waste, the dangerous,
And all to sweetness turns.'

From *Digging* by Edward Thomas

In order to develop a deeper understanding of smells and of how the mind perceives them, I journeyed deep into a wood in rural Dorset. I drove far up a country lane that had taken me miles from the main road, its hedges thick with hazel, hawthorn, ash and sloe. The further I went, the thicker the hedges got, and the narrower the lane, and for the last mile I could barely squeeze through. My destination was a large wood of oak, hazel and ash trees, where I was to meet Jim

Bettle, a charcoal burner and the head of The Dorset Charcoal Company. Jim was delayed, and it was Sam and Mike, who do the hard work for Jim, who let me through the forestry gate and drove me to the clearing a mile into the wood. Here we found piles of old oak logs, a corrugated iron shack for tools and chainsaw oil, and three great cylindrical metal kilns. One kiln had a scarf of thin smoke drifting from it, colouring the woodland air pale blue in columns of light and dark where the sun fell through the trees. 'Nice office,' they said as we drew up in the clearing, 'no noise, no people.'

'Not over-bad,' said Sam, inspecting the kiln. 'That kiln's about ready, you can tell by the smoke.' The kiln, about four feet high and six feet across, had a metal lid shaped like a Chinese hat, shimmering with heat, and three or four narrow chimneys set into the base, from which the thin smoke oozed. Close by were two more kilns. One had its lid off, showing that it had been loaded with freshly split oak logs ready to be burned. There was a metal pipe down the middle, which Sam explained was to allow a fire to be dropped into the centre to start the charcoal burning. The other had its lid on, and had been cooling for a day: it was ready to be opened and for the charcoal inside to be bagged up.

So straight away I had three delightful smells to savour. The smoke from the first kiln, which had almost finished burning, gave the warm, sweet, savoury aroma of burning oak: something close to roasting meat but also with vanilla and an edge of bitter tar. From far off, it was gentle and sweet, but close up it was cloying, acrid and tarry. The second kiln, still unlit, smelled of the delicate natural odour of oak, which was far from the smell of resinous softwoods, more like a blend of sour milk and tanned leather. The third kiln, burnt through and cooled, smelled of cold ash – dry, tarry and a little ticklish to the nose. Three different smells with some common features, all variations on a theme of oak, all giving a sensation of comfort, all intensely pleasurable, all charged with memories of other times when the smell of wood smoke or fresh split oak had been in the air.

Here we have the first quality of the sense of smell that sets it apart from the other senses. Its first message is an emotional one: 'I like this smell; I dislike that smell; I am suspicious of this other smell.' Before my little expedition to Dorset, I had interviewed Professor Thomas Hummel, who runs the Taste and Smell Clinic at the University of Dresden. He is one of the few experts who understands the sense of smell in a holistic way, not just as chemistry or physiology or psychology, but as a complex interaction of all these. 'There's a very intimate connection between the olfactory system and emotions,' he told me:

> 'As soon as we perceive an odour, we like it or we dislike it. It always has an emotional quality. With sights or sounds, there isn't always that emotional quality. We see a house on a hill or we hear a dog barking in the dark – much of the time we perceive these things without emotion. But we smell rotten eggs, and everyone has an immediate emotional reaction. That is very probably because, inside the brain, the olfactory bulb and the amygdala (the small brain organ that is responsible for much of our emotional activity) and hippocampus (another crucial brain organ that is important for fixing long-term memories) are all together, with only a synapse or so separating them. And it is a good thing – it's

> the task of a smell to make us do something, and the emotion is the whip.'

While I sniffed around, Sam had been preparing the second kiln for lighting. He filled a small brazier with chips of half-burnt wood and fresh shavings, and lit it with the help of some diesel fuel. Now he jumped on top of the unlit kiln full of oak logs, removed the central chimney and was ready to pour fire down the centre of the pile. He explained how it works:

> 'We load about one and half tons of oak into the kiln, and we expect to get a quarter ton of charcoal out. We lay the wood so the air is channelled in from the ports low down. That gets it going. When it's maybe three quarters alight, we put the lid on, propped up at first to let the air in fast, with the metal chimneys on top, so they are warmed up and draw properly. Then we fix the lid down, seal the gaps to prevent air getting in, put in the chimneys to draw off the smoke, and let it burn overnight. About eight hours in, the oils in the wood will have formed a gas ball in the kiln which makes for an even burn all round the inside of the kiln … Next day, I look at the smoke and seal off the ports when it's looking right. Then it takes a full day to cool off.'

Before Sam lights the kiln, we have a conversation about the different smells of timber. There is plenty of fresh split oak to smell, most of it English oak but there are also a few logs of Turkey oak, and some of Scots pine. They are all delicate smells, difficult to hold in the mind and to put into words.

There is, to my surprise, not one simple wood smell, but a great variety of woods. We are familiar with the resinous aroma of pine and other softwoods, but few are familiar with the smells of hardwoods, which are quite different. Later, when I paid a visit to a specialist timber yard where many kinds of hardwood are stored, I received a lesson in the nature of woods which suggested that there are very different chemical signatures for different trees. The chemical signature of each tree runs all the way through the aroma of their fruit, their leaves, wood and roots. So there is timber that smells meaty (larch or bog oak); fruity (pear or apple, naturally); creamy (lime); nutty (walnut); malty (yew); of chloroform (iroko) or of vanilla (beech). Deep in that Dorset wood, without recourse to chemical knowledge, we struggled to describe the logs that were to hand.

Sam admitted that he was at a loss to describe the smell of fresh oak. After a lot of sniffing and a long pause he ventured: 'Smells like new furniture. I can't manage better than that. It keeps slipping away from me.' He had hit on another particular feature of the sense of smell. It occupies a position on the very edge of our attention. It is not at the centre of our attention like vision and hearing, nor is it completely below our attention, like our sense of touch or of balance. There is an obvious reason for this, and a deeper, hidden reason. The obvious, practical reason is that smelling is tied to breathing. It is as short lived as an in-breath. We breathe in; the smelly molecules flood up our nose; our brains register them; we breathe out. On an out-breath we smell nothing (unless we are eating). The next breath will probably bring a subtle change of odour to which our brains may have to react, maybe a change that signals danger. So we have very little time to think about the successive smells that come to us as we breathe.

The hidden reason for our sense of smell being just out of the centre of our attention is suggested by the rare experience of people who have temporarily exaggerated perceptions of smell. It is worth taking a break from the charcoal to consider this condition, which is called hyperosmia, because it may explain why the sense of smell occupies the position that it does in our minds. Oliver Sacks, the neurologist who wrote about people with unusual brain conditions in *The Man Who Mistook His Wife for a Hat*, and other brilliant books, described a medical student who experimented with drugs, and developed hyperosmia. Although Sacks does not say so explicitly, it seems likely that he was describing his own experience.

One night this medical student dreamed that he was a dog, living in a world of vivid and significant smells. When he woke up, he found that his sensitivity to smells really had been transformed, that he really was in a world of vivid and significant smells. Overnight he had become hyperosmic, someone with an abnormal and exaggerated response to smells. For the few weeks that this hyperosmia lasted, smell became his dominant sense. Every street, every shop, every person that he encountered had a smell that held its own signature for him, more telling than the sights of that street, that shop or that person. The pleasure of nice smells was more pleasurable, and the revulsion of bad smells was more revolting.

It seemed to him not so much that his world had more intense olfactory sensations, but that his whole world had to be understood in a different way. It was a world of overwhelmingly immediate and concrete significance, so much so that he found it hard to reflect and to reason in abstract terms. Giving such primacy to his sense of smell interfered with thinking, that activity so important to human beings.

After a few weeks, his hyperosmia faded out. His sense of smell returned to normal. He was back in the normal world, a less vivid world than the one where smells meant so much, a disappointingly faint, and muted world. But he was back in a world where he could resume the mindset of a thinking, reflective person.

What had happened to shift the sense of smell from its usual position – low on this student's mental horizon – to a position of such dominance? Nothing had changed in the structures of his nose and brain. He had not grown the snout of a dog with its interior convolutions of bone. He had not developed a new set of olfactory nerves overnight, nor had a host of new connections been created in the deeper parts of his brain. Presumably what had happened was not a sudden new growth but the removal of some restraint, some inhibitor, which normally keeps the sense of smell from being too much the centre of our attention. An overpowering hyperosmia like this is very rare. Many women during pregnancy experience something that is described as hyperosmia, although it turns out to be slightly different.[11] It is not so totally overwhelming as the hyperosmia described by Oliver Sacks, more a selective change in their emotional response to certain smells which could be harmful to the growing baby.

However rare this overpowering form of hyperosmia, the fact that it can materialise overnight is very suggestive. What it suggests is that we do not lack the equipment to respond like a dog to the world of smells. When the inhibitors are removed, we sniff almost as purposefully as any dog (almost – but a dog can sniff eight times per second with ease, whereas humans can only manage four sniffs per second with an effort), the smells are at the centre of our attention, and we linger over the meaning of the scents like any dog. But there is a price: if the world of smells occupies more of our attention, we find that we have less attention to spare for thinking and reflecting in abstract terms. Humans are thinking creatures who have carved out our niche in the world using our capacity to think, to organise and to communicate. So this case of temporary hyperosmia suggests that in most of us our sense of smell is normally damped down, in order to leave us mental space for other brainwork. It occupies, as a result, a strange position on our mental horizons.

Sam and Mike, the charcoal burners, had been busy. They had taken the lid off the kiln that had cooled, and started to bag up the contents.

> 'You can tell when it's good charcoal by the sound. It's like tinkling glass when you break it up. We bag up the bigger bits for restaurants. The smaller bits are for home barbecues. The smallest bits go to the yard to be milled into powder. You'd be surprised what the charcoal powder is used for. There are charcoal biscuits, and charcoal soap, and charcoal in face powder and toothpaste. There's charcoal in our first aid kit: it's like a filter if you have swallowed poison. We sell a lot for horses. Did you know that horses are one of the few animals that can't vomit? So they need food supplements containing charcoal to take away the toxins. Especially retired racehorses that build up lots of toxins inside as they lose their muscle. Charcoal's the best filter.'

Sam moved from kiln to kiln as the work demanded, never idle. His next task was to light the kiln full of fresh oak ready for burning. He climbed on to the woodpile, poured the fire into the hole left at the centre, and listened for the gentle roar of air being sucked in by the fire, then for the growing crackle of timber catching fire. He capped the central hole with hefty logs, placed the massive rusty metal lid on top of the kiln, and propped it with logs between the lid and the metal cylinder. The fire was beginning to take hold in the depths of the kiln, heating the sides, making the rusty metal shimmer with heat, and sending great gouts of smoke out through the gap between kiln and lid.

On and off the kiln, in and out of the smoke, Sam was as smoked as a kipper. I told him that the smoke must have tanned him through and through; he himself would be inured to the smoke, but the smell would be the signature by which he was recognised. He replied that he had long since ceased to notice the smell, except when he got his work clothes out in the morning. He had become habituated to the smell long since. Here is another special feature of the sense of smell: we adapt to repeated smells faster than to any other repeated sensation. It is anybody's guess why this should be, but a plausible reason is that we need our sense of smell to stay alert to new dangers and olfactory signals, and it benefits us therefore to blank out repeated smells. It takes much less time to cease noticing a smell than to blank out a repetitive sound or sight. You can demonstrate this for yourself. Take a bath with an unfamiliar scent in the air, and time how long it takes before you have ceased to notice the sensation of heat in the water, the sight of the taps, the sound of the water cylinder refilling, and the smell. The chances are that your sensitivity to smell will have faded after four or five breaths. Your sensitivity to heat will have faded more slowly, disappearing where your skin was completely immersed but remaining alert for some time at the surface where your skin sensed the change in temperature between water and air. Your sensitivity to an unchanging noise will probably have faded more slowly, and your sensitivity to sight will probably have remained unchanged.

We habituate to a repeated smell in a few breaths. We adapt long term to a prolonged smell, blanking it out completely in a few hours. The blanking out is more complete than with the other senses, and smell also takes longer to return to normal sensitivity when the prolonged smell is no longer there. Without that capacity to adapt, it would be difficult to recruit for some of the smellier jobs. Adaptation is odour-specific. If you work in a rubber factory, your nose will selectively tune out rubber, but it will remain sensitive to roses, roast beef or rosemary. The selective nature of adaptation is sometimes exploited by perfumers when they try to match one fragrance to another. A perfumer will use what is called 'saturation sniffing' to become so used to a scent that any differences between the original scent and the scent that is intended as a match will stand out. The perfumer will repeatedly sniff the target scent, so that his brain is totally adapted to it, and he then sniffs the matching scent: his brain will filter out any sign of the original scent, and any remaining minor differences will then stand out. Adaptation is a useful feature of any sensory system: it preserves our ability to detect small differences between stimuli against enormous variation in overall intensity, and so allows us to catch the faintest whiff of a menacing smell such as smoke.

The newly lit kiln was by now well alight. Swirls of smoke were curling from under the propped-up lid and up into the sunlit wood. The metal cylinder of the kiln was alive with heat, its rusty surfaces were shimmering and distorting the surrounding countryside, and a steady low thunder of fire was roaring inside. Jim, the head of The Dorset Charcoal Company, had driven up in a truck, accompanied by an enthusiastic young dog. Jim told me about the economics of charcoal:

> 'Like all country businesses, the margins in charcoal are so slim that you're always one cock-up away from going bust. But it's important to have a home-grown charcoal business. That way we help to give English woodland a sustainable economic future. And it's so much less damaging than imported charcoal, some of which is a by-product of the permanent devastation of forest. It's environmental rape.'

Meanwhile Sam and Mike had finished emptying the cooled kiln of charcoal. They had time for a chat about smells. Jim's dog attracted attention to himself, picking up logs in his slobbery jaws and offering them to all the company. Mike said:

> 'Dogs, now, they know about smelling. Always smelling each others' little calling cards. The foxhounds sometimes come through the wood here when they are being exercised. Must be 50 or 60 of them, all with their noses to the ground. Wonderful sight. They would be able to tell you about the smells of the wood, if they could talk, mind.'

Mike, having finished his work, began to talk lyrically about the smells of the woodland. His companions folded their arms, leaned back with the detached look of men who thought that lyricism about smells was not going to be Mike's strong card:

> 'There's lovely smells in the wood. Bracken, wild garlic, lovely smells. Brambles, they smell. Sounds daft, but I like the smell of ash logs – that's a proper smell like, that's the proper log smell. My favourite time is when it's wet. The damp makes everything smell more. Like yesterday, the rain was falling in the wood, birds were trilling, you could smell the growth.'

'Yesterday,' said Jim, in a flattening tone of voice, 'you sat in the truck out of the rain and listened to the radio. Fat lot of smelling you did.' It was time for me to be off.

CHAPTER 5

The Meanings of Smell

The business of detecting smells happens in the nose; the business of *recognising* smells happens in the brain. We must follow the trail of smell into our brains to discover how we recognise smells, remember them and think about them. It is a 'hot' trail, as we might say of a scent that a bloodhound is following, one that has been recently laid down. Much of the science of olfaction in the brain has been uncovered only relatively recently, and is still being uncovered. It shows that the sense of smell is unusual in several ways – less hard-wired at birth than other senses; more tailor-made for us individually and less off-the-shelf; more revealing about hidden meanings and less informative about obvious ones; more closely connected to emotion and to long-term memory than our other senses.

The first relay station in the brain for information coming from the receptor nerves in the nose is the olfactory bulb – found just the other side of the skull behind our eyebrows. This small brain organ receives information from the nerves in the nose and organises it for transmission to other parts of the brain. Whereas receptor nerves are organised to pick a smell apart by responding to different types of molecule, the olfactory bulb is organised to put that information back into a pattern. It is not the individual molecules of a whiff of coffee or fresh baked bread or frying bacon that we recognise, but the overall smell of coffee; not the 27 detectable molecules given off by a cup of coffee, but the overall pattern made by those molecules. Our brains have registered that pattern before, and they rapidly synthesise the information coming from the latest whiff to match the pattern already in our brains. Inside the olfactory bulb, then, we start putting the fragmentary information about molecules back into the pattern that means 'coffee' or 'baked bread' or 'frying bacon'. The nerves from the nose converge in their thousands on a clumped set of brain neurones called glomeruli, in the outer layer of the olfactory bulb, which in turn have multiple connections to just 25 or so big cells (called mitral cells) that transmit signals to the rest of the brain. Millions of receptor nerves thus converge on a very small number of transmitters, allowing a mess of sensation to become a recognisable pattern. A gas chromatograph does the job in an hour or so; our brains do the job in a fraction of a second.

The route to the brain taken by information about smells is quite different to the routes taken by our other senses. The olfactory bulb, the first point of call of our sense of smell, is located within easy reach of some important brain structures that are at the heart of our emotions and our long-term memory. The other senses take a longer way round, through a brain organ called the thalamus which serves as the relay station for all other sensory information apart from smell. In the heart of your brain, beneath the thick convoluted grey matter, is what is known as the limbic system. This is a collection of structures which support such essential brain-work as emotions, behaviour, motivation and long-term memory.

The olfactory bulb, which transmits information about the sense of smell, is part of the limbic system, and so the sense of smell is more intimately connected to emotion and memory than our other senses. This very close connection between the olfactory bulb and the limbic system probably explains the unusually evocative character of smells. It probably explains, too, why our immediate response to smells is to like or to be disgusted by them, as no other sense has such privileged access to the emotions.

In the early days of brain science, around a century ago, the limbic system was described by anatomists as the 'smell brain' because in many creatures (including humans, as noted above) the sense of smell is located within the limbic system. It was also sometimes called the 'snake brain': in snakes, and in other creatures that have little capacity to think, but a big capacity to smell, the limbic system is pretty much all that their brains consist of. It has also been called 'the ancient mammalian brain'. The notion that the human sense of smell is primitive was reinforced by these observations that, in creatures with much less complex brains than ours, smell occupied such a prominent place. Nowadays, these simple ways of thinking about the limbic system have been abandoned. Brain scientists now understand that the wiring of the brain is so interconnected that it is misleading to isolate different activities within its particular structures. Perception, memory, emotion, thought and responses to sensations activate many parts of the brain across the boundaries of the limbic system and the grey matter of the cortex. Scientists recognise too that the sense of smell in humans is not such a primitive activity; that it takes up a surprisingly large chunk of our genetic code; and that it has sophisticated interactions with other senses in our brains.

Amongst the many connections reaching outwards from the olfactory bulb into the rest of the brain, there is one that seems particularly important. This is the connection that leads into the cerebral cortex – the layer of grey, infolded matter that most of us would conjure up if asked to visualise a brain. It is here that much of the activity of thought takes place, and a particular fold of grey matter called the piriform cortex does much of the thinking about smells. The piriform cortex has two major divisions with anatomically distinct organisations and functions. The front part is associated with determining the chemical structure of the odorant molecules, while the back division is known for its strong role in categorising odours and assessing similarities between them. It is here in the back division for instance that we spot the mintiness of mint chocolate and menthol cigarettes, or the smokiness of kippers and cigars; the oakiness of oak-smoked ham and of wine stored in oak barrels; or the metallic quality of blood and garlic.

A pair of US scientists won the Nobel Prize in 2004 for unpicking the way in which the human genome carries the blueprint for the sense of smell. They are Dr Linda Buck and Dr Richard Axel.[12] Their work, originally published in 1991, opened up the science of olfaction as never before. Since that pioneering work, Dr Axel has led research at the University of Columbia on how smells are represented in the brain. He says that looking inside the brain at a microscopic level, we find some fundamental differences between smell and the other senses. All of our senses have to perform a similar trick in order to translate the real world from sensations of real things into the electrical language of the brain's neurones. The colours and sounds and smells and kicks from the external world do not enter our brains, they are represented electrically –

i.e. turned into an electrical pattern. Most of our senses do this representation by preserving something of the pattern that hits our sensory nerves in the deeper parts of the brain that do the interpreting, recognising, understanding and remembering. So in vision, for instance, there are brain cells in the visual cortex that are tuned to particular features in vision, such as vertical lines, horizontal lines or lines at an angle. And brain cells that are tuned to such features are clustered together. The brain is to some extent ready-wired to interpret vision, and the wiring inside the deeper parts of the brain bears some resemblance spatially or temporally to the wiring of the receptor nerves. There are similar hard-wired features in other senses, except smell.

In the sense of smell, the receptor nerves in our noses are activated by smelly molecules, and communicate their messages to the olfactory bulb just behind our eyebrows. So far so good: the impulses arriving in the olfactory bulb are wired in a predictable way, so that one set of receptor nerves communicates with one part of the olfactory bulb. You could, with a very powerful microscope, look into the olfactory bulb and read which neurones are activated, and (with knowledge of the patterns) you could say which smell was being received in the olfactory bulb. From the olfactory bulb, there are communications with many parts of the brain, including the piriform cortex, a thinking part of the brain that learns and recognises smells.

Inside the piriform cortex, however, there are no 'mother' cells that are pretuned to particular smells. There are no predictable patterns, mirroring the input from the olfactory bulb. There are no clusters of cells responding to a cluster of features of a smell. Nothing appears to be hard-wired.

A smell is represented by an ensemble of neurones which is not pre-determined. The same smell in one person will be represented in another person by an entirely different ensemble of neurones in an entirely different part of the piriform cortex. If you forgot the smell through some accident, the chances are that you would relearn the same smell with a different clutch of neurones in a different part of the cortex. It is a blank slate.

The Columbia team's work, particularly on mice, has demonstrated the random nature of the brain's handling of smell. They have, for instance, been able to compare in fine detail the location of smells in the brains of different mice, and the fact that there is no common organisation in the mouse's piriform cortex. They have, in another instance, been able to wipe out the representation of a smell from the piriform cortex of a mouse and recreate it in another location in the piriform cortex. They have taken a smell to which mice are highly averse, like fox urine, and relocated it while keeping the mouse's aversive reaction to the smell, and they have also turned the mouse's understanding of the smell of fox urine from aversive to attractive. In mice the brain is not hard-wired for smell; it is a blank slate on which each individual mouse's experience of smells is written in their particular way. It is the same for us humans.

Apart from a small number of odours which elicit innate responses, the vast majority of odours are recognised in the piriform cortex in a random way. It is an apparently disordered system, but the order is imposed by our personal experience. It is, says Dr Axel, a system which allows you to respond to an unpredictable and variable world by learning.

This explains many important things. It explains, for starters, that our sense of smell is the product of our personal experience of life, from our mothers' breast onwards. Newborn babies show very few inborn smell preferences. They show a slight tendency to wrinkle their noses at bitter smells and at strong cheesy smells. They are not likely to persist with such preferences through life, as they learn the pleasure to be had from bitter drinks like coffee and from ripe, runny cheeses (the flavours of which, as we shall see, are more a matter of smell than of taste). The absence of inborn responses to smell must, we can guess, be an evolutionary advantage to creatures like us who are omnivores. Because humans, like pigs and rats, will eat almost anything, we benefit from being able to adapt to many different diets in many different parts of the world.

If our sense of smell is a blank slate at birth, that explains why it is such a variable sense from person to person. Each of us learns the meaning of smell from experience, and our experiences are different. Trials have shown that one person's sensitivity to a smell may easily vary from another person's by a factor of 1,000.[13] People who are very sensitive to one odour may be quite insensitive to another. Many of us have olfactory blind spots, odours that others perceive and we do not. These blind spots are called 'specific anosmias', and it seems that a high proportion of us have specific anosmias. A recent study of 1,600 volunteers found that one in three was anosmic to at least one of the twenty odours they were asked to detect. For some of the odours tested, particularly those like sandalwood with a high molecular weight, as many as one in five of the volunteers were smell-blind. Compare this to vision, in which deficiencies are much less varied: the number of receptor types in vision is small, and as a consequence the number of forms of colour blindness is limited. It is natural that those senses that we use for orienting ourselves in space – vision and hearing – should have less variability from person to person.

If we could not be sure that everyone was seeing the world in the same way, we would be in trouble when, for instance, crossing a street or negotiating a path through a crowd. The fact that much of the mechanism for these senses is inborn helps us to survive, even if they still have to develop further in the first years of life. In humans however the sense of smell is much more personal, concerned with discriminating our personal world and our personal experiences and our personal meals, and for these purposes it does not matter that the sense of smell is variable from person to person.

No one teaches us the meaning of smells. Parents teach their small children the meanings of the things they see and hear with persistent and loving instruction. They encourage their children to spot a tractor or a car transporter as they drive on a family outing; they carefully teach their children to form syllables and words, and to string them into meaningful sentences; but no one helps us distinguish the smell of coffee from coconut, or straw from strawberries. We teach ourselves and we are limited to what we experience. So the meanings of smells are our own personal meanings, coated with the associations that we have made through our personal experience, and charged with the emotions that we have felt.

It is small wonder, then, that there are large cultural differences. Smell being so much the product of our experience, it is natural that people brought up in cultures different from ours should have very different preferences and associations for smells. So you find studies comparing, for instance, the way German and Japanese people perceive certain smells. Germans love the smell of marzipan; Japanese people think it smells of oil or sawdust. Japanese people are more tolerant of the smells of dried fish, cheese and beer; German people are more tolerant of church incense, aniseed and almond. Japanese people refer to Westerners, apparently with no offence intended, as '*bata kusai*' – smelling of butter.

For the cattle-raising Dassanetch of Ethiopia, the odour of cows is beautiful. Men wash their hands in cattle urine, and smear their bodies with manure. Women rub themselves with butter. The Dogon of Mali find the scent of onion the most attractive fragrance. For the Bushmen of the Kalahari, the loveliest fragrance is rain. No doubt the Dassanetch, Dogon and Bushmen would be dumbfounded by the smell preferences of Western Europe. They would, I guess, boggle at our taste for smells that are sweet almost to the point of nausea, like strawberries, and to perfumes that are almost as aggressive as petrol.

Although there is great variation from person to person, there are some general rules. Typically, old people are more prone to losing their sense of smell. One in four people over the age of 50 has significantly decreased olfactory function; one in three people aged over 70 has effectively lost their sense of smell and derives no pleasure from the flavour of food. That is not surprising, given that our nerves for olfaction renew themselves so fast and so frequently: in old age, many of us are bound to reach the limits of our capacity for renewal.

Women are typically better at discerning and recognising smells than men. However, this is not true of children: boys typically have olfactory abilities as good as girls'. But grown women are typically better than grown men. Professor Hummel, of the Taste and Smell Clinic in Dresden, has interesting things to say about why there should be this gender difference:

> 'When we test olfactory function, women are typically much better than men. That is true for all aspects of

olfactory function – for threshold (perceiving odours at very low doses); for discrimination, and for identifying an odour. Also for odour memory: there's a game called "*Smellery*" using odour memory – it's terrible to play with women. They always win.

'Perversely, it does not seem to be because women have better systems for smelling. The olfactory bulbs of men are typically larger than for women, though there is some evidence that the olfactory bulbs of women are more densely packed. No, it's my guess that women are better because they are more interested. Olfactory function is very much affected by interest. If you are interested in smells, you perceive smells better. Odours are a social signal, and women are better socially, so they are more interested in the social meanings of odours. That's just my hypothesis, but I find the same thing every day in my clinic. The men come to the clinic because their wives send them, and they complain only that they can't smell the red wine. The women with olfactory problems come because they are really upset by the social side.'

Professor Hummel would not claim that social communication is the only function of our sense of smell. There are other important functions – detecting danger, recognising familiar and unfamiliar territory, savouring flavour in food and drink. But the social significance of smells is of abiding interest.

It is body odour that conveys much of the social meaning; body odours come from sweat and from the bacteria that live in our sweat glands. Human beings are sweatier than many other creatures. We have more sweat glands per square inch of our skin, especially those that produce 'emotional sweat' – the sort that pricks the palms of our hands and the soles of our feet when we are anxious, frightened or in pain. Not only do we sweat under emotional stress, but our sweat is distinctive – our individual smells are a personal identifier, almost as unique as a fingerprint. There has been serious research into the possibility that machines might identify people by their individual body odour, for instance at immigration controls. That research has not yet led anywhere, because there is no machine capable of unpicking smells as reliably and as fast as noses can. Sniffer dogs, however, can identify people's individual smell from sweat samples. The only people who pose a problem for sniffer dogs are identical twins. When they are young, before environmental factors alter the smells of identical twins, their body odours cannot be distinguished by sniffer dogs. That is a clue to the fact that genetic factors are an important ingredient in body odour.

The most important social communication through smell comes about because of genetic factors, because human beings can recognise blood relatives by smell. This is a skill that women have much more than men, as Professor Hummel asserts, and it is a skill that is at its peak when women are ready for childbearing. It has been known for some time that mothers can reliably identify their own biological children by smell, and pre-adolescent siblings can recognise their siblings by smell.[14] This finding was interestingly expanded by some recent research at Wayne State University in Detroit, where studies were carried out into families with step-siblings and half-siblings as well as biological siblings. In these families, mothers could reliably identify their biological children by smell but not the other children. Siblings could usually identify their birth siblings by smell, but not their half-siblings or step-siblings. It all suggests that the sense of smell is active in creating the bond between mothers and babies, and indeed the pleasure that mothers take in their babies' smell is good evidence of that bond. We see the role of smell in bonding in many other animals as well as human beings. Sheep, for instance, rely on smell so much for bonding that they can be fooled by smell into adopting a lamb which has been spurned by its mother: if a ewe has given birth to a dead lamb, the shepherd will quickly skin the dead lamb, and wrap the skin around a lamb which needs to be adopted. The ewe sniffs the skin of its own offspring that has been transferred to the orphan lamb, and bonds with the orphan lamb as if it were her own.

There is a wealth of research into other forms of social communication through body odour. Fear, aggression and lust are sweaty emotions, and many research projects have tried to investigate whether the sweat from these emotions is communicative. On the whole, the evidence from such research is flimsy. There have been, for instance, many trials in which people are exposed to terrifying situations, and their sweat is sampled for others to sniff. People jumping out of aeroplanes for a first parachute jump, and people watching horror films are favourite subjects. Pads from their armpits are given to other subjects to sniff and to say what they can detect. These trials tend to show that 'terror sweat' does not communicate strongly enough to evoke a conscious reaction, but it can produce an unconscious reaction that shows up in brain scans as increased activity in areas of the brain concerned with social signals and the emotions of other people. Such trials tend to show too that people are not reliably able to tell the difference between sweat produced by terror and sweat produced by vigorous exercise.

As for love, lust and sexual attraction, these emotions are of such interest that they will be dealt with at greater length later in this book. Suffice it to say now that smell does play a role in these emotions. But it is not the role that most of us would expect.

There is a specialist form of communication through odour whose significance has dwindled over recent centuries, but which remains important. This is the odour of illness. Before the 19th century, doctors would routinely sniff their patients as part of diagnosis, and midwives would sniff their way through the stages of childbirth. Although such medical practice seems like quackery today, it is still a fact that some illnesses reveal themselves through smell. The sweat of people with schizophrenia has a component called trans-3-methyl-2 hexenoic acid, with a scent that is described as very sweet like over-ripe fruit. Cystic fibrosis patients have breath that is said to be slightly acidic, as do asthma sufferers. Untreated diabetics suffering the advanced stages of hyperglycaemia often have breath that smells like lemon drops or rotten apples. Liver and kidney diseases can make breath smell fishy. There is even a disorder called 'stinking fish syndrome'. Diphtheria has a sweet smell. Typhoid patients smell of baking bread, and a standard diagnostic test for all kinds of diseases, very much in use today, is for doctors to smell the patient's urine.

Tuberculosis gives off a smell like stale beer. It is still a virulent killer, accounting for well over a million deaths a year. In Africa, a squad of pouched rats (African Pouched Rats are a much more likeable breed than our sewer rats) has been trained to sniff out tuberculosis. They are not only much faster than laboratory tests at detecting TB, they are also more accurate. They are acting under the orders of Bart Weetjens, a Dutch man who found a happy coincidence between his love of rodents and his desire to make the world a better place. His squad of rats began their careers sniffing out landmines, at which they were in every way superior to people with mechanical mine detectors: fast, fearless and not liable to trigger mines accidentally.

For midwives, too, smells are important. In the antenatal clinic, they are on the lookout for smells that may indicate that a fœtus has been exposed to nicotine, alcohol or drugs, or smells that indicate that the mother's metabolism may be under stress. Midwives can tell the progress of a birth by smell. The onset of labour may be announced by the smell of amniotic fluid leak. If there has been an undetected leak and infection has developed, there will be a foul dustbin smell of bacteria. Once the baby is born, its smell may also be important diagnostically, as several metabolic disorders cause a change in the smell of skin and urine. These are conditions that do not show up until birth, but need to be caught immediately a baby has been born if seizures and brain damage are to be prevented. 'Maple syrup urine disease' is a disorder caused by an enzyme defect in which the baby's urine smells of burnt sugar. Similarly, in the metabolic disorder called phenylketonuria, the baby's skin and hair gives off a musty or mousy smell, whilst isovaleric acidaemia causes an odour reminiscent of sweaty feet. Tyrosinaemia gives the baby a smell of boiled cabbage.

Birth has its smells, and so does death. We have the evidence (admittedly oblique and unverifiable) of Oscar the cat.[15] Oscar was adopted by a nursing home for people with advanced dementia in the USA, and spent most of his time in the unit for people approaching the end. Staff noticed after a while that Oscar would make his rounds, just like the doctors and nurses. He would sniff and observe the patients, then curl up to sleep on the beds of certain of them. The patients on whose beds he curled up would often die several hours after his arrival. Oscar's attendance at deaths became so regular that staff took it as a predictor that a patient would die within the next six hours. It moved from being a regular coincidence to a firm predictor on the occasion that a doctor pronounced that a patient would die soon, and Oscar stalked out of the room. The patient did indeed last longer than the doctor had predicted; Oscar returned to the room later and the patient duly died a few hours after his return. Did Oscar smell approaching death, or was he responding to some other hidden signal?

Professor Hummel of Dresden says that many of his patients who have suffered permanent loss of the sense of smell also suffer from depression. He quotes a female patient: 'When I wake up at night, it's just black. I have lost all the little clues that give texture to the night.' For first hand evidence, I went in search of someone who had lost her sense of smell. I found that the loss is not as dire for everyone who suffers it as Professor Hummel asserts. I met a young woman who had been knocked over by a car years before. In the accident she had fractured her skull, and had been left with partial hearing loss and the total and permanent loss of her sense of smell and her sense of taste. The accident had jolted the cribriform plate through which the nerves of smell pass into the brain, and had severed the lot. The accident happened when she was eighteen; now in her thirties with two children, she had adapted thoroughly to the damage done to her senses. She found it hard to remember what it was like to smell and taste things. The smell of her babies, a visit to a perfumery in Provence, an aromatherapy session – these were pleasant ideas but gave her no sensation at all. 'I have a memory of a memory of the smell of some things – like the smell of dogs or the beach, but I don't feel I'm missing anything. Memories are what you create.'

At the time of the accident she had been studying for the hotel and restaurant trade. 'That seemed like a bad choice when I lost my sense of taste.' Food now meant little to her, though she did most of the cooking for the family, and relied on them to guide her on matters of flavour. For her own palate, she looked for variations in texture to give her food some interest. Though she missed the clues that smell would have given her about her environment, she found herself sensitive to the feel of air, its humidity and thickness. 'When my six year old was sick the other day, I went into his room and sensed something was wrong. The air seemed thicker somehow, like there was petrol vapour or a heaviness in it.'

Perhaps the biggest effect of losing her sense of smell was her underlying compulsion to check for leaks or fires or dirt that she might otherwise have detected by smell. 'I'm obsessive about airing things. I spray all the time. I probably make the children wash more than they need. I make everyone brush their teeth a lot.' On two occasions since the accident she thought she smelled burnt toast: once, it was a phantom smell, but once there really was toast smouldering.

For her, the loss of the sense of smell (and the loss of flavour that went with it) was regrettable but not a source of sadness. She was admirably positive about life and was not going to let this loss interfere with it. I read to her a case study of a man who had lost his sense of smell, from *The Man Who Mistook His Wife for a Hat*. She pulled a face at what seemed to her an unnecessarily maudlin response to losing the sense. Sacks describes a man who sustained a head injury, severely damaging his olfactory tracts, who in consequence entirely lost his sense of smell.

He had not given his sense of smell much thought before he lost it. But when he did lose it, he found that his whole world was suddenly radically poorer. Smells, he now realised, had given him the very groundbass of life, and without them he felt an acute sense of loss and an acute sense of yearning.

The young woman I met had no such sentimental nostalgia – no 'veritable osmalgia', as Sacks puts it – for her sense of smell. She told me that 'of all the senses, smell is the one I would most happily do without.'

For those of us who have retained our sense of smell, we are hardly aware how much it gives texture to our lives, how it forms, in Oliver Sacks's words 'the very ground bass of life'. We often struggle to dig out the meanings of smells, but there are meanings and they come in many different forms. There is the meaning that comes when we recognise the substance that gives off that particular smell. There is the meaning that comes when we perceive that that particular smell is like other smells from other substances. There is the meaning that comes when a smell sparks a memory. There is the meaning that comes when a smell tells us of a familiar person or a familiar place. There are social meanings as smells tell us about the nature and feelings of people we encounter.

While it takes brain scans and microscopes to investigate how the neurones in our brains handle a smell perception, we can find out a lot without technology, simply by observing how we process odours. It takes only a little practice and reflection to observe what we do with a smell once it has registered in our noses. Here are seven observations I have made about how we process odours.

My first observation is that we recognise the pattern of a smell in a single act of recognition. We can unpick a few of the components in the pattern, but that comes later. The first thing is the 'Aah, Bisto!' moment. Often that is as far as we get: we can easily recognise the smell of coffee, but it takes much more effort to discern that this particular whiff of coffee has a smokey or a vanilla component. We often get as far only as the 'tip-of-our-nose' effect, the sensation that we should be able to recognise the aroma but it is just beyond our reach.

My second observation is that we recognise both complex and simple patterns of molecules in much the same way. A whiff of coffee, with its many molecules, is as easily recognised as a whiff of grapefruit, with a simpler pattern of molecules. There are however limits to the complexity that our brains can handle. Research chemists working on synthetic fragrances or flavourings explore those limits on a daily basis. They are often commissioned to come up with the simplest facsimile of a smell like coffee. It has to be simple, because that is cheaper and more easily controlled for quality, and it has to be lifelike. (More insidiously, there is an incentive to create a synthetic smell that is just slightly brighter and sweeter than the natural smell, because that will appeal to the consumer's exceptionally sweet tooth, a product of centuries of Caribbean sugar, and that will encourage them to prefer the brand smell to the real thing.) This process gives some clues about how much olfactory complexity our brains can handle.

In recreating the smell of roasted Arabica coffee, for instance, these chemists might begin by selecting, from amongst the 27 odorant molecules that have been detected by the gas chromatograph, only those molecules that are present in concentrations well above our sensory threshold. If a blend of these selected molecules does not match the original smell, they extend the list to include odorants at or below the sensory threshold. Once a blend matches the full smell closely, it is tested further. One by one the aromatic molecules – the odorants – are subtracted from the formula. If the resulting formula smells less realistic, the molecule that was subtracted is put back; if the subtraction makes no difference, that odorant is dropped. The end product is the mix that most resembles the original with the smallest number of ingredients. In the case of roasted Arabica coffee, a passable synthetic copy of the smell was made with sixteen different kinds of molecule, out of the original 27. That suggests that our brains can handle a mix of about sixteen components, and will detect that a smaller mix is unrealistic, but that our brains cannot tell the difference between a mix of sixteen components and a more complex mix.

We have a limited basic lexicon of recognisable smells in our heads (Observation 3). No doubt the lexicon varies radically from person to person, and is determined by the experiences they have had and their command of words. No doubt as well, the lexicon can be extended by training and new experiences. But it is limited, and it is full of gaps. In my own case, I can recognise a few score smells, certainly less than 100. Many of the items in my lexicon come from the kitchen cupboard – the various spices, herbs and fruits I keep there. In my vocabulary of basic smells there are relatively few items from nature – I cannot distinguish the smell of one kind of grass from another, nor one kind of earth from another, and I would not rely on my sense of smell to distinguish one kind of animal from another.

In theory, a combinatorial system like smell, where a recognisable smell can be made from any permutation of a large number of variables, should offer us a very large number of smells. We have 352 different kinds of smell receptor, and we can cope with permutations of up to roughly sixteen variables, so the theoretical possibilities could be a whopping two with 27 zeroes behind it – roughly as many permutations as there are atoms in the human body. So it is surprising that we recognise so small a number of smells in real life. However, the flexibility of our sense of smell is demonstrated by the next observation.

Although we have only a limited vocabulary of basic odours in our heads, we are able to discern very many subtle variations around the basic smells that we recognise (Observation 4). With practice, we are capable of naming hundreds of variations in the bouquet of a glass of red wine, or a glass of whisky. We can happily walk around a scented garden and dwell on the many variants we find in the fragrance of roses or geraniums or mints. We are capable of spending many happy minutes sniffing the different cheeses in a well-stocked cheese shop. People who work with coffee or tea or cigars distinguish scores of attributes in the aromas of those products. It is an intriguing thought that we might develop as sensitive a discrimination of other things – apples or bread or cotton – were they as expensive and worthy of attention by connoisseurs as wine and whisky are.

My fifth observation is that we can recognise some components in a smell much more easily than others. Smokiness, mintiness, citrus, anise, sweatiness, chocolate and coconut, amongst other odours are easily discerned as components of more complex smells. Earthiness, grassiness, the peculiar tang of tomato stems, woodiness, dampness are qualities that announce themselves in an aroma much more quietly. One might have expected that the most common smells would be the most easily recognised, but that is not the case. It is rare for me to encounter coconut and very common for me to encounter grass, but the rarer coconut is much more detectable than the more common grass. No doubt this anomaly is explained by the relative power of the molecules conveying these smells.

My next observation is that we have no difficulty distinguishing the main smell from the subsidiary smells. We have no difficulty, for instance, when we encounter a mint-flavoured chocolate or a chocolate-scented mint flower in knowing that one is a sweet and the other a flower, without seeing or touching either.

Finally, I observe that we can discern an enormous palette of sweet, sour, bitter, sharp and other abstract qualities in an aroma. In this respect, our sense of smell is quite different to our sense of taste. On our tongues, the taste of sweetness is one, unvaried thing; it is the same sweetness whether it derives from a lump of sugar or an orange ice lolly. In our noses, the sweetness of roses is different to the sweetness of oranges or the sweetness of hay or of incense. In our noses, the qualities of sweetness can co-exist with other qualities like smokiness or bitterness or piquancy, and give an entirely distinctive sensation.

The sense of smell, then, is adapted to our particular environments and shaped by our personal experiences. It is not a sense, like vision and hearing, which conveys communal meanings, meanings we share with other people and which we can readily communicate to others. It tells us simple facts about the presence of different odours in our environment; it does not tell us, as vision and hearing do, where things are, how big, how fast-moving, how intense. So our sense of smell combines good absolute sensitivity with good quality discrimination. Good absolute sensitivity means that for some odours we spot their presence in minute doses: for many odours, the nose is still the most sensitive instrument we have. Good quality discrimination means that our noses are very good at distinguishing minute differences. Molecules that are so similar to each other that they have the same constituent atoms in the same quantity and the same order but with a different orientation smell different to our noses. In fact, it is much easier to detect that two successively presented smells are not the same than to detect that they are identical. In vision, the reverse is true, and identical images are detected much faster by the eye than images with slight differences. Hence those games in which two almost identical pictures are shown, and you have to spot the ten minor changes from one picture to the next. At the same time, our noses are weak at discriminating intensity: it takes an increase in the concentration of a smell by 20 per cent before we perceive a difference. In vision, an increase of 2 per cent in brightness is enough for a difference to be detected.

Brilliant at detecting tiny doses of smells in the air; weak at spotting changes of intensity. Brilliant at distinguishing minute variations in smells; weak at telling what they are or where they come from. Powerfully formed by experience and coloured by emotions – all this is indicative of a sense whose function is to tell us about the familiar and unfamiliar in our environment, and to make us feel at ease in our environment, rather than, as with vision and hearing, interpreting the meaning of the world.

CHAPTER 6

The Sense of the Familiar

Long distance travel is disorientating. You begin the journey at an airport terminal with a bland, characterless, international flavour, and you end the journey in an airport terminal that is almost identical. The signs at your destination may be in an alien lettering, but there is a familiar English translation to make the international traveller feel reassured. The passport control may be conducted by officials of an alien government, but they are dressed in familiar prison-warder uniforms, well-creased with jangling metal bits. The architecture of the terminal building is international glass and steel. The advertisements are for familiar airport international brands. The trolleys are the same as back home. The car hire stalls are the same; the coffee shops are the same; the escalators are the same.

You may have travelled to the other side of the world, to Tokyo Haneda airport for instance, but you do not feel sure that you are abroad. Outside the airport terminal, there are still disorienting sensations of sameness. The sparrows are the same and make the same chirping sounds as back home. The tarmac is the same; the same Japanese makes of car on the road are the same as back home; the grass is the same (well, almost the same), the anonymity of public transport is the same; the hoardings are for the same brands. It is only when you get out of your taxi or train, and set foot on the streets, that it hits you that you are abroad. And the sense that brings it home to you is the sense of smell, the smell of foreign cooking – in the case of Tokyo: the smell of stalls selling hot soba noodles or yakitori – or the smells of foreign rubbish and foreign smoke and foreign cleaning fluids.

There is nothing like a foreign smell to tell us that we are really abroad. There is nothing like a homely smell to reassure us that we are home, or a homely smell in an unhomely place to bring on homesickness. More than any other sense, smell tells us what is familiar and what is unfamiliar. It gives us our sense of place, of being at ease with our environment or on alien ground.

Writers often use smells as a powerful way of evoking the character of a place. D.H. Lawrence, writing of Mexico, said that each country has a faint, physical scent of its own, as each human being has. Mexico's scent, in his view, was compounded of resin, coffee, perspiration, sunburned earth and urine. Baudelaire, in notes for a travel book that he never wrote, was brisker: Paris for him smelled of old cabbages, Cape Town of sheep, Russia of leather, Brussels of coal tar soap, and the Orient smelled of musk and corpses. Evelyn Waugh summed up Athens by its smell of garlic and dust.

Like many other animals, human beings make strong associations between places and their smells. Home is full of our smells and they are reassuring. Homesickness – the discomfort of being in unfamiliar places without home comforts – is often provoked by smells. For a child, sent away from home to hospital or an aunt's house or a boarding school, it is the smell of these alien places that is at the heart of the misery they evoke. The medical smells of hospital, compounded of methylated spirits and unhomely detergents; the disturbing smell of an aunt's unfamiliar cooking and cleaning habits; cheap floor polish and cheap disinfectants in a school – these smells tug at the heart-strings. There is a reverse homesickness too, provoked by catching a whiff of home smells when you are away from home, that reminds you of the comforts of home and of the people you miss. In *The Wind in the Willows*, domestic old Mole is coaxed out of his cosy hole to experience the wider world. The outside world is a vivid place, with river trips and Toad Hall and caravan expeditions. But a moment comes when Mole catches a fleeting whiff: 'Mole … stopped dead in his tracks, his nose searching hither and thither in his efforts to recapture the fine filament, the telegraphic current, that had so strongly moved him … Home!'

Smells, as Kipling said, are surer than sights or sounds to make the heart-strings crack.

Geographers have recently cottoned on to smell as a factor in geographical thinking. 'Urban smellscapes' are a newly coined device for mapping the smells of a city. The late Victoria Henshaw, the originator of these urban smellscapes in Britain, saw their purpose as being to understand how smells contribute to our sense of place, with the practical aims of reducing the nuisance of obnoxious smells and of preserving delightful smells. She started from the perception that odours are a very significant part of our experience of cities, and that they have meanings. Too much of our response to urban smells, she believed, is about pollution, poor air quality and 'nuisance' smells. Urban planners and retailers have been led, by that negative view, to eliminate smells, and to create a bland environment. Her view was more positive: that we should also look at the opportunities to savour the characteristic smells of places, and to allow cities to express their nature through smells that give pleasure and tell us about the people who live there and their activities. She developed 'smell walks' as a valuable mechanism both for documenting odours that are specific to places, and in order to bring out their hidden meanings.

One of the founding ideas of smell geography derives from Japan, a country with a powerful respect for tradition, and a willingness to keep tradition alive for the sake of it, not for primarily economic reasons. Japan has an official list, sponsored by their Ministry of the Environment, of 'One Hundred Places of Good Fragrance'.[16] The list serves to identify places worth visiting for their fragrance and worth protecting. It includes the scents of liquor and soy sauce from Kurayoshi warehouses; grilled sweetfish from the Gogasegawa River; Japanese beeches and dogtooth violets on Mount Kenashigasen; the vapour of Beppu's eight sulphur springs; the scent of the rocky coast at Iwami Tatamigaura; the early morning market at Hida-Takayama; and the streets of used book shops in Tokyo's Kanda district.

How endearing of the Japanese, amidst the hurly burly of manufacturing electronic gadgets and cars and manga games, to give time to so quiet and unproductive an activity as sniffing. It is hard to imagine a Ministry of the Environment in a more materialist country thinking it a sufficiently hard-headed task to sponsor a list of 'One Hundred Places of Good Fragrance' in its own country. I, however, think that every country would benefit from one.

The British list of fragrant places should, no doubt, be determined by public competition and decided by a panel of worthies. I hope the list will include a good number of places that smell of smoke. Wood smoke from salmon and kipper and eel smokeries. Roasting coffee in the many little roasteries to be found in every city. Coal and steam smells from preserved railways, so evocative of the glum post-war world before the 1960s, or from steam fairs. Charcoal burning in hidden clearings in the old forests which yield the oak and hazel wood needed for its production. Braziers in city streets at Christmas time where chestnuts are roasted.

Then our list of fragrant places should include places where the air is rich with the smell of fermentation. Traditional breweries where the fermenting of barley with hops and yeast creates a daily sequence of odours as the beer is brewed and the residues are taken to create pig swill or Marmite (a by-product of brewery yeast). Silage pits where hay is crushed and mixed with molasses to create animal feed. Cider presses, perry presses and wineries where fruit juices are fermented, and where, best of all, the leftover skins and pips create a cheesy compost. Dairies where cheese is made, another product of fermentation.

No doubt many people will propose herb gardens and rose gardens for our list of good-smelling places. Certainly some must be included, but only those which offer a special experience. Not the usual limited selection of well-known herbs, but the few places that offer a complete spectrum of those plant families that come in a mind-boggling variety of fragrances. Mints, for instance, which give odours that range from cat's piss to sponge cake; or scented geraniums which offer odours of absinthe, brilliantine or coconut.

Our list of good British smells is far from exhausted. There should be a place for good marine smells – places which smell of seaweed or marine mud or tidal marsh. There should be a place for the friendlier animal smells – the aroma of horses for instance in some horse-racing town, of bran and leather and dung and dust. There should surely be a place too for the fragrance of trees; of lime trees in blossom or of pine forest or of hardwood timber yards or of old libraries.

What a thing it would be if we could reverse the trend towards blandness, and actively celebrate the pleasures of great British smells. We could have railway stations in Kent smelling of hop sacks; in Somerset of apples; in Northumberland of kippers; in Rutland of Stilton cheese; in Birmingham of metal filings. We could have aromatic tours of redolent places. We could have aromatic parades by fleets of mobile tarring units, farm muck spreaders, steam rollers, hay wagons and shire horses.

One of my personal 'places of good fragrance' is the suburban town where I was born and spent some of my childhood. I decided to make an olfactory expedition there in search of the smells of my childhood. It was a commuter town in the semi-countryside of Surrey, south of London. I went back there in search of the smell of the familiar, of things that would carry the powerful emotional charge of nostalgia. I went in search too of the essence of suburbia – of a town of small Victorian villas, of tennis courts, of municipal offices, of tile-hung pubs.

A short walk from the railway station I encountered the first characteristic smell of suburbia. From the front hedges of a line of Victorian villas came the scent of common privet in flower – a quiet, unassumingly sweet scent. Privet is the emblem of English suburban front gardens – the plant of privacy which forms part of the outer defences of the Englishman's castle. In mid July it bears an unobtrusive panicle of white flowers hidden amongst the well-clipped leaves. It has a scent of the most modest sweetness, a half-hint of vanilla blended with the flat smell of brown wrapping paper. Such quiet, unemotional smells evoke the soft-spoken, polite, unexciting nature of suburban streets.

The fragrance of privet sparked in me an unsought nostalgia for my adolescence in this town. It brought back for me not only the gentle scent of privet hedges of my past, but also in the same breath the sensation of being an unmotivated teenager mooching along the warm pavements of the town in the summer holidays, not knowing what to do nor how I might fit in with this unexciting world. I felt again the self-conscious awkwardness of a fifteen year old with nothing to do.

And indeed as I wandered the town I noticed several of the same strange breed to which I had once belonged. A pale youth was sitting by the station on a low box of telephone equipment, cradling his face in one hand and looking poetically into the distance. In the High Street I saw a teenage boy with an extravagant hairdo following his mother into a shoe shop and trying to look as if he had no connection with her. On the other side of the street there was a teenage girl with a polite demeanour, exaggeratedly short shorts and a wild jacket with too many zips and buckles. All through the town these specimens of awkward adolescence were to be seen amongst the shoppers, and I felt for them.

Privet is not the only hedging plant in suburbia. It is in the nature of the small villa garden to have an unshowy individuality, to demonstrate its difference from its neighbours without going too far. So a back street in this Surrey town is a succession of varied front hedges with varied smells, all within polite boundaries. Laurel has a glossy, leathery leaf that gives off a delightful whiff of green bananas and almond. Holly and ivy smell bitter. Box leaves give a warm homely scent of dry wood and tobacco. St John's wort with its bright yellow flowers has an unexpectedly vivid smell of lemon drops and sesame. Cypress Leylandii, the invasive monster of suburban England, gives off an invigorating blend of turpentiney resin and citrus. Aucuba, a hideous shrub with yellow blotched leaves and evil red berries, gives off a blend of meaty leather and greenery. Hawthorn has leaves that smell of ripe apple and flowers that smell of an awkward blend of honey and fish paste.

As I walked on savouring the individuality of suburban houses and their gardens and hedges, I was surprised to see many small fig trees that had seeded themselves amongst the hedges and in the brick-and-flint walls. They must, I thought, be a relic of Victorian plantings of a relatively exotic tree. I plucked and sniffed a leaf. One whiff was enough to ring a loud bell in my memory. In my childhood garden there was an old fig tree espaliered against a side wall of the house, and the smell of the fig leaf now chimed instantly with the smell of fig leaf then. A dry, warm, cardboardy smell, with the rich fruit scent that goes with prunes and (of course) figs; a smell that has an affinity with other warm, dry things like autumn bracken or straw or dry muesli.

The memory of childhood came to me in a way that is peculiar to the sense of smell, and unlike memories evoked by sight, sound or taste. The memory came instantly, with no need for mental enquiry. It came unsought. It came trailing other sensations, not only olfactory sensations, but also the sensation of being small, the sight of that particular wall with those particular orange-pink bricks studded with wall nails, the kinaesthetic sensation of pausing in my childish rush past that side wall. It is my impression that, although few of us have the synaesthesia which means that our senses run together, for most of us olfactory memories are synaesthetic in nature and arouse memories in other senses as well as in the sense of smell.

I walked on past more detached houses with their individual gardens. Here was one with Mediterranean pretensions, planted with lavender and santolina and a tiny olive tree. Here was one with *folie de grandeur* – a concrete eagle on each gatepost, a brand new bought-complete garden and an expensive car. Here was an endearingly shabby garden with no pretensions – its car had flat tyres and twining brambles, and its weather-beaten wooden fence had gaps. A wooden fence, treated long ago with creosote, is a delight to the nose: a comfortable dry resinous scent with hints of tar and of the earthy mushroom smell of decayed wood. Each front garden has its individual character; each house has its stylistic idiosyncrasies; each combination of front gate, front path and front door is different; each house has its individual name eccentrically different to the names of its neighbours. And, because these suburban streets were new in Victorian times, each house bears the imprint of a dozen different owners since then and a score of different fashions in gardening and house decoration. It is the gentle cult of individualism and homeliness enshrined.

Whether intentionally or not, the owners of suburban gardens are marking their territory with the mild smells of front gardens, and giving their families an olfactory navigation mark. I doubt, though, that many gardeners give as much attention to the fragrance of their plantings as to the looks. They will plant a honeysuckle and rose primarily because it looks good, and be pleased that it also smells good. If fragrance were the main factor, then flowers would count for less than leaves. For the little-known fact is that the leaves of plants offer much more variety and beauty of fragrance than their flowers.

After the back streets, I wanted to savour the High Street and its shops: an expedition that became a survey of what I considered the honest versus the dishonest smells of shops. I was not expecting the High Street to offer nostalgic smells. For one thing, in my childhood many shops smelled of unwrapped food, of meat, baking, fruit and vegetables. Today almost all food is wrapped and refrigerated or chilled, and gives no smell. For another thing, all those shops that baffled me in childhood with their medieval titles – drapers, milliners, ironmongers, haberdashers, grocers – have gone. But I was to find that there is plenty of character in the smell of shops: some honest and some conveying a false sense of synthetic sweetness or freshness or cleanliness.

First stop was a tobacconist. There was no smell of tobacco, just the saccharine scent of sweeties. Here was a thoroughly dishonest smell, too sweet to come from nature; it did not convey chocolate or strawberry or coffee or liquorice, just synthetic sweetness. Cigarettes and cigars were for sale, but were hidden behind a metal screen. In a very obscure corner of the shop there were tins of tobacco sold for pipes or roll-your-own cigarettes. I was not allowed to sniff them, but I could see from the labels that most of these tins contained synthetic, even dishonest versions of tobacco – rum and plum; spearmint; lemon and honey; sherry. Pipe tobacco is often treated with 'casings' or 'toppings' to enhance the natural flavour of the leaf. A casing is a liquid marinade, absorbed by the leaf during curing, sometimes a solution of sugar or chocolate or liquorice. At the more vulgar end, toppings are aromatic supplements added to the final blend which burn off quickly. The trend towards synthetic smokes continues into e-cigarettes, where the majority of liquids that smokers 'vape' have aromas that are far removed from tobacco.

Next was a charity shop, one of many in the High Street, selling used books, used clothes, used shoes, used DVDs. Here was an honest smell, albeit one of the saddest smells. Dusty, musty, mousey, old, the smell of old wardrobes in old houses where fading aunts and uncles have died. The bookshelves where I browsed were sad too, books from the recent past whose passing will never be regretted. I found books about hobbies of the past – model making and flower arranging – books about forgotten sportsmen, books by forgotten gardening experts, books about diets that have had their day. The smell of ageing books and ageing people, and the knowledge that both the books and their owners were dead, blended into one melancholy experience, the smell and the objects bound together in feelings of sadness.

After the sadness of charity shops, I was invigorated by a genuine breath of fresh air – a florist. What makes a florist smell so fresh? It is a mix of the heady scent of lilies and peonies with the complex menthol of eucalyptus leaves and a deep green freshness given off by the buckets and vases containing water and flower stems. It was so floral a smell that I suspected an added chemical. But no, I was assured by the brisk woman with rolled-up sleeves, there was nothing added.

Next I encountered a succession of shops dominated by the stink of rubber. A bike shop smelled of tyres with an undertone of mineral oil. A shop selling fishing tackle stank of thigh-length rubber waders. A sports shop smelled of trainers and tennis balls. I struggle to put words to the smell of rubber. It is a smell that stands alone, with no obvious points of comparison. To my nose it is a cloyingly offensive smell. But apparently some pregnant women develop an appetite for the smell of rubber that has to be satisfied by a manic search for rubber gloves or the tyres of children's bikes.

The smell of rubber has been analysed through a gas chromatograph. It reveals a scent made up of a large number of strong odorants from several classes of compounds. There is not a single dominant molecule, but an ugly blaring chord of them that smell of tarry smoke, fish, carbolic and overheated plastic.

Next I came to a gift shop. It was a pretty shop with pretty pastel products for sale, a pretty pastel girl at the till, and a strong fragrance. But something was not right. The fragrance was too fragrant. It came from a display of scented candles, each candle charming in its way, but each one excessive. There were candles smelling of vanilla and summer berries, cinnamon and honey and sandalwood. But natural smells do not come in such an unrelenting flow at such intensity, with so little contrast. It was too perfect; it was the sweet smell of excess. It is an unnatural effect that we encounter through other senses as well as the sense of smell. We see computer-generated images that are too bright, too intense, that have too little plainness to offset the over-stimulating colours. We hear computer-generated music that is too perfect, that has no edgy notes to offset the perfection, that has a bass that is too throbbingly deep. We taste flavours that are too sweet or too savoury. Our sense of touch too is told lies, with fabrics that are too satiny or too fluffy. Even our kinaesthetic sense is told lies as we experience fake work-outs on machines in gyms. The synthetic spoils our appetite for the natural by overfeeding it.

It was a relief to come at last to a shop which smelled honest, with no phoney fragrance. It was a hardware shop. Almost everything in its stock had a genuine smell that had not been tampered with or overpackaged. There was linseed oil, teak oil, lubricating oil, Danish oil, tung oil. There was paint, fertiliser, grass seed, bird seed, charcoal in sacks, brown wrapping paper, boots. There was the glorious smell of ropes made of sisal, hemp and jute, and broom heads made of wood and bristle. It was a potpourri of varied scents making a complex blend of mineral, vegetable and animal. Even outside the entrance it gave off an arresting scent.

Inside I found two middle aged men wearing old-fashioned brown storemen's coats. I had to explain that I was not intending to buy anything, but only to savour the smells of their shop. They exchanged glances which said 'here's something odd', and started a comic routine between themselves. 'I think Sir will appreciate essence of wellington boot,' said one. 'Sir won't be looking for essence of wellington boot,' said the other: 'Sir will be after something altogether more suave. Perhaps Sir will find fragrance of coir doormat expresses his masculinity.' 'Or perhaps Linseed Oil Number Five,' said the first. 'No doubt Sir is into *Eau de* Glue,' said the second, 'many of our more sophisticated clients are into *Eau de* Glue.' At this point, the first bloke started an astonishing routine, based no doubt on TV adverts for perfumes. He strutted down the shop, each step exactly in line with the previous step, waving an imaginary chiffon scarf at tins, bottles, boots and packets as he passed, and wafting their imaginary fragrances under his nose. His mate and I watched in astonishment. At the far end of the shop, next to the rolls of rope, he twirled to a finish and gave us a sultry look. We fell about in amazed laughter.

After that show, we had a more sober discussion of the smells of the shop. We agreed that the smells of rope deserve a particular word of praise. Modern ropes are made of polypropylene which has merits of durability but not of smell. But a self-respecting hardware shop also sells ropes made from traditional fibres. Jute, the Bengali fibre that gives us burlap and hessian, can be made into a rope that smells meaty and leathery. Sisal, the fibre of the agave plant, gives us baler twine and dartboards and sash cords; it smells of bitumen. Hemp, the fibre of the cannabis plant, gives us the origin of the word canvas because sails were once woven from it, and its smell evokes marquees and country shows. Because hemp sails tended to rot when long exposed to sea water, hemp also gives us the expression 'Jack Tar' for sailors, since canvas had to be tarred. Were I a perfumer, concocting my ideal scent, I would want the essence of these fibres as the bass notes for my perfume.

After the High Street, my next destination was the local park, a ridge of sand and sandstone overgrown by scrubby oak and fir trees, with a few fine beech and chestnut trees. It was regarded by my late father as a minefield, where he was sure to step accidentally in dog muck. At the entrance there was a notice which would have fuelled his anxieties – addressed to 'Responsible Dog Owners'. Not many of those, to judge by the evidence on the sandy paths.

Dog muck was only a minor element in the prevailing odour. Leaf mould, giving an old, dry smell compounded of sour tannin and the tobacco scent of old leaf and the earthiness of fungus, was one major element. The other strong scent instantly opened a door onto childhood – the scent of bracken. At first sniff this was a smell so completely recognisable that there was no call for further analysis. It was the unmistakable, characteristic, distinctive, indescribable smell of bracken – brackeny.

With an effort and some false starts, I managed to ask myself – is it a rich, complex smell or a simple smell? The answer came as I rubbed a frond between my fingers: it was a smell that evolved over time, starting as a simple smell of fresh vegetation like grass, and then developing a deep, rich fruity note close to the scent of figs, and finishing with an autumnal scent of leaf. This trick of avoiding the first impressions when recognising a smell has become for me an essential step in appreciating them. All those scents that hit you with instant recognition – be they vinegar or cigar or tar – and are immediately 'indescribable' or 'unmistakable' or 'distinctive', can be described if you can sidestep the instant recognition.

At the top of the little ridge that formed the park, there stood a wooden rotunda. The awkward teenagers of the locality had left their messages of angst scratched into the wood of the rotunda's central column – 'Fred woz ere innit blud', 'will I or wont I', 'aw it's hard being middle class'. My expedition had taken me back in time to the characteristic smells of an English commuter town. It had evoked with nostalgic intensity my childhood through the modest scent of flowering privet, the brackeny smell of fig tree and the figgy smell of bracken, the sadness of charity shops and the delight of an aromatic hardware shop. It had reminded me of the hidden connections between varied things whose scents ring the same bells in the brain – the connections that link the smell of bracken and fig and brown wrapping paper and suburban pavement.

The smell of a place is remembered deep in our brains, and it will last a lifetime. Stir up a smell, and the memory will return in vivid detail, more vivid than memories stirred up by old photographs or old recordings. Smell memories are different to other memories, and the emotions that go with them are different to the emotions that go with other senses. So what exactly is special about the memories and the emotions of smell?

CHAPTER 7

Madeleines

Smell is emotional and evocative. So, of course, are all the senses. It is essential to us that our senses should trigger emotions – the emotions in their turn trigger a reaction to what our senses perceive, whether the reaction of self-defence in response to fear, or the reaction of flight in response to the emotion of disgust, or the urge to eat in response to a pleasurable sight or smell of food. It is also essential to us that our senses should stir up memories – without the memories we would be as befuddled as babies by the meaningless sensations that bombard us. It is memory that gives meaning to our senses, by filling in the gaps of what we cannot discern, or by enabling us to interpret as complete something of which we can only discern a fraction.

Smell is however *more* emotional and *more* evocative than our other senses. Not only is it more emotional, but the emotions that it stirs up are a peculiar lot. And the connection between emotions and smells works both ways – emotions affect our perception of smells and smells affect our emotions. This two-way conversation between the emotional and the olfactory is not to be found with our other senses – we do not, for instance, lose some of our vision when we are unhappy, but we do lose some of our olfactory power when we are unhappy.

Smell memories, too, seem to be less under the control of our consciousness and more likely to take us by surprise. For most of us, it is not possible to memorise a smell in the way we can memorise a word or fact, and it is not possible to bring a memory to mind in the way we can bring a fact to mind. We acquire smell memories by accident, without conscious effort, often without being at all aware that we are smelling anything. And then, years later, the memory is brought back to life, again without conscious effort and often without being aware that we are smelling anything, and the memory brings the distant past to vivid life.

Psychologists call this the 'madeleine effect' in honour of Marcel Proust's great book *À La Recherche du Temps Perdu*, in which he captured the special power of smell memories. He certainly did not discover the effect, but he fixed it for all time in a single image of a madeleine, a tiny aromatic French cake dipped in tea, which inspires the author. The first hundred pages of this book are a delightful fictionalised memoir of Proust's cosseted life as the only child of indulgent, well-heeled parents living in a French country town. Proust dwells on memories from every sense, and tries to bring them to full life in words. The look, the feel, the texture, the taste, the smell, the emotional attributes of everything in his childhood are brought back into being. There is hardly any story for a hundred pages, just wonderful description.

In the family house in the fictional French town of Combray lives Aunt Léonie, who gave up an active life years before and retired to her rooms with an unknown, unnamed affliction. She lives a life of low activity but high drama, attended by Françoise her maid who indulges and scolds her in equal measure. Much of Aunt Léonie's time is spent fussing about her health, with frequent disapproving observations through her window of the movement of neighbours. On Sunday mornings Aunt Léonie invites her little nephew into her bedroom and gives him a little madeleine – a tiny sponge cake shaped like a scallop shell – dipped in her herbal tea:

> And suddenly the memory returns. The taste was that of the little crumb of madeleine which on Sunday mornings at Combray (because on those mornings I did not go out before church-time), when I went to say good day to her in her bedroom, my aunt Léonie used to give me, dipping it first in her own cup of real or of lime-flower tea. The sight of the little madeleine had recalled nothing to my mind before I tasted it; perhaps because I had so often seen such things in the interval, without tasting them, on the trays in pastry-cooks' windows, that their image had dissociated itself from those Combray days to take its place among others more recent; perhaps because of those memories, so long abandoned and put out of mind, nothing now survived, everything was scattered; the forms of things, including that of the little scallop-shell of pastry, so richly sensual under its severe, religious folds, were either obliterated or had been so long dormant as to have lost the power of expansion which would have allowed them to resume their place in my consciousness. But when from a long-distant past nothing subsists, after the people are dead, after the things are broken and scattered, still, alone, more fragile, but with more vitality, more unsubstantial, more persistent, more faithful, the smell and taste of things remain poised a long time,

> like souls, ready to remind us, waiting and hoping for their moment, amid the ruins of all the rest; and bear unfaltering, in the tiny and almost impalpable drop of their essence, the vast structure of recollection.

This tiny moment of genteel life has become the symbol of the power of olfactory memory. It does not detract from its significance to find that Proust tried some other symbols before lighting on the madeleine. His notebooks containing early drafts of the book show him first having a bite of honey on toasted bread dipped in herb tea. A second draft has the memory sparked by a 'biscotto' – a hard, nutty Italian version of a biscuit. Only at the third take did he settle for the madeleine, distinctive enough to spark so strong a memory, common enough to convey a simple meaning, genteel enough to evoke fastidious bourgeois life.

Smell memories are more long-lasting and less prone to being edited in our heads than memories of sights or sounds.[17] There is another quality that smell memories have, one lightly touched on by Proust, but one whose significance deserves to be teased out. A smell memory activates all the senses – it is synaesthetic. Synaesthesia is the happy condition in which people have a sensation in one sense that triggers a sensation in a different sense as well. So people with synaesthesia may find that a sound has a colour, or a shape has a sound, or a day of the week gives them a sensation of a particular body position. The forms of synaesthesia are very varied, and people can experience several forms together – the painter Kandinsky for instance had a four-way synaesthesia involving colour, sound, shape and taste. The connections between sensations are often arbitrary – one person who sees letters in colour will always see an 'A' as a pale brick red while another will always see an A as green. The connections are fixed from an early age and do not change over time – if A is pale brick red it remains so for the rest of that person's life. It seems a rare and peculiar condition, but the latest estimate from the British Association for Synaesthesia is that one person in every 27 may have the condition in one form or another, often without realising it.

Synaesthesia involving smell is comparatively rare, and for the few people who have synaesthesia involving smell, it may take a great variety of forms. To one such person, the smell of coffee might have the look of grey triangles. To another, it might have the sound of an electric fan. Some people have synaesthesia which links smell to feel ('the smell of celery is like sharp points on the arm'). For others the link is between smell and concepts ('the smell of bacon is like the concept of hope') or between smell and personalities ('the smell of cherries is like a ten-year-old boy') or between smell and an emotion ('the smell of chocolate is nervous'). Then there are the reverse synaesthesias, in which a sound, a touch, a concept might evoke a smell ('the 1700s smell like daffodils').

One theory to explain synaesthesias is that, in the infant brain, the boundaries between the senses are not distinct, that there are many connections within the brain between the various senses, and that we shed many of the connections that are unimportant to us as we develop. So the senses separate themselves in our brains, and most people are left without these interesting but unimportant sensory harmonies in their heads. In two ways, however, the sense of smell continues to be connected to other senses for most people. One is in the illusion of flavour, which we will shortly see is constructed by a combination of the senses of smell, taste, feel, sight and sound. The other function of our brains in which the sense of smell activates other senses is smell memories. It is possible that everyone is a synaesthete when it comes to smell memories.

Take the memory provoked by the smell of fresh rain. I step outside when a summer shower is falling, and I am transported back in time by the damp earthy fragrance to the moment when this memory was laid down. I am transported instantly, with no effort of memory on my part. The memory evoked by that fragrance is not just of a smell, but it is also the sensation of being a small child, of the sight of raindrops falling in a suburban gutter, of the feel of the clothes I wore as a child. More than that, it is a memory of the emotional quality of being a child, my awareness of the adults with me and what they might say about my being out in the rain, my fear of some of the looming houses in the street, my attraction towards the park gate at the end of the street, and my childish shock at the rude graffiti scrawled on the gate. None of these sensations would have been evoked by the same memory conveyed by another sense; a photo of myself as a child in a shower of summer rain, for instance, would not have carried a fraction of the multiple sensations conveyed by the smell.

Our heads are full of the smells of childhood. As we age, the smell memories endure. By the end of our lives, our heads are full of extinct smells. My head is full of the extinct smells of my childhood, and the heads of the next generations will be full of smells from their childhoods that in their turn will have disappeared from ordinary life, and will be preserved only in memory. The smell of steam trains was a feature of my childhood and has all but disappeared. It lives on in my memory, along with the rubbery steam of car radiators that had boiled over on a taxing hill, and of ageing leather from car interiors.

There are other extinct smells – the harsh aroma of French and Balkan cigarettes; the sad stink of boiled cabbage in grim lodgings; the eggy stench of fluid for cleaning silver and of ladies' hair remover; fishy smells of bad electrical wiring, given off by an ancient plastic called Bakelite. In my head are preserved also the cheery smells of coal tar soap and shampoo, and the fresh sourness of vinegar used for rinsing hair; the meaty aroma of ropes made from natural fibres; dried apple smells and fruit cake from that extinct domestic treasure trove, the larder; the school smells of desks and pencil boxes, glue and spilt ink. The old smells of dentists are now extinct – a clove scent to disguise the foul stink of the dentist's drill (that smelled like burned hair) and the menacing edge of chloroform. Industrial smells linger on in my head from the age of the factory: the cold smoke stink of coking plants and gas works; the oily onion stink of metal factories, and the rich oiliness of oil remover; ash and cinders.

These extinct smells seem to be ancient history but they endure in the head of the older generation. As for the children of today, who knows which of the smells they experience now will be preserved in their memories long after they have disappeared from everyday use? Pipe and cigar smells have almost disappeared, and the aroma of cigarette smoke will surely follow. Many food smells will become unfamiliar as more of it is sealed in airproof containers. Some smelly technologies are bound to disappear, so that, for instance, the smell of petrol exhaust or gas cookers will probably become extinct within the lifetime of today's young people. Anxiety about pollution and safety may drive out the aromas of bonfires and barbecues, as it has already driven out the late summer aroma of burning stubble from cornfields. Furniture smells are changing as fashions and materials change, and the smell of polish or wool may become extinct. Children's toys create some of the most enduring smell memories, and some of them are sure to go, so that an ageing generation will preserve the treasured memory of Play-Doh or electric trains when they are no longer sold.

Given the particular power of smell memories, it is not surprising that nostalgia is one of the emotions most easily touched off by smells. No one who, as a child, had to spend unhappy nights away from home will forget the alien smells that went hand in hand with homesickness. The reverse effect can be just as strong – when unhappily away from home we catch a whiff of a home smell, and our unhappiness is brought back to the surface by that smell of familiar cooking, or of clean laundry, or of homely polish.

Homesickness is a neglected emotion. Up to the 20th century it was seen as a serious ailment, and it was claimed that people died of it. 'Hypochondria of the heart' was considered a disease for which there was only one cure – to return home. In 1688 we find a Swiss doctor studying the homesickness of Swiss mercenaries serving away from home, and inventing a new term for the malady – 'nostalgia': from the Greek words meaning pain for homecoming. In a recent study (*Homesickness: An American History*) Susan Matt combed through medical records from the American Civil War, and found that nostalgia was listed quite frequently as a reason for soldiers being discharged, and sometimes as a cause of death. By the end of that war more than 5,000 soldiers had received a diagnosis of nostalgia, and 74 Union soldiers had died of it. By contrast, the US records of death in the First World War list just one soldier as having died of nostalgia, and it appears that no one has died of it since.

It seems likely however that homesickness continues to flourish, continues to cause acute unhappiness, and continues to be fed by the sense of smell, but that psychologists now put the malady under the more general label of depression. It would be interesting to investigate how extensive and how severe homesickness is amongst the people swept away from home in the great waves of migration that have followed every war of recent history. It would be interesting too to find out how the migrants create a sense of home with familiar smells in their new locations, and how those smells, familiar to immigrants but alien to the native population, contribute to the occasional waves of hostility to migrants.

It is a curious feature of the intimate relationship between smells and emotions that our perception of smells is altered by our emotional state, and, vice versa, that our emotional state can be altered by smells. Take, for starters, disgust. Newborn babies show little response to smell, but at a very rancid or a very bitter smell they do wrinkle their noses and stick out their tongues, making the same face of disgust that adults make. The disgust response is universal, and it is easy to elicit; in fact it may be easier to elicit disgust than almost any other emotion. Things that provoke disgust are obvious contaminants – blood, vomit, faeces, rotten flesh, pus – but also things that are associated indirectly with contamination like foreigners, deviant acts, and people who are perceived as living dirty lives.

Some people are easier to disgust than others. And their sensitivity to disgusting smells appears to go hand in hand with more severe political attitudes. So people who are instantly disgusted by, say, the smell of rotten flesh are likely to have harsh views on immigrants or sexual deviance. More interestingly, people's political attitudes can be changed by smell.[18]

Put people in a room with an evil smell, and it is likely that their political views will shift markedly towards the harsh end of the spectrum. So views on homosexuality for instance, are likely to harden when the ambient smell is rotten egg. There could be a new role for 'ambient scenting' (the new vogue for using fragrances to influence people's behaviour and attitudes) here. A political party with an intolerant agenda could, for instance, drum up more vigorous support by pumping out ugly odours to accompany rabble-rousing speeches. The reverse may be true as well: that pleasant smells may encourage more charitable attitudes. Perhaps such is the unintended function of incense in religious ceremonies.

The connection between bad smells and harsh emotions goes in both directions. Bad emotions can turn smells bad, in much the same way that bad smells can turn emotions bad. A piece of research at the University of Wisconsin-Madison in 2013 tested people's reactions to smells in situations of high anxiety. As well as the usual, subjective methods for assessing people's reactions to smells by subjective ratings, this research used MRI scans to observe the subjects' brains at work, providing a more objective view of what happens within the brain when we smell something. The research subjects were exposed to disturbing images and descriptions of car crashes and war, and their reactions to a series of smells were tested. Under that stress, they perceived as unpleasant smells which they had previously rated as neutral. Anxiety had heightened their sense of disgust.

More striking still, your ability to detect smells and analyse them is affected by your general mental condition. In this respect, the sense of smell differs from all the other senses: you do not lose your ability to see, or hear, or feel if you are unhappy, but your ability to smell is markedly diminished when you are unhappy.

Depression, schizophrenia and seasonal affective disorder all suppress the sense of smell. To find out why, researchers at the University of Dresden Medical School exposed people – 21 with major depression and 21 who were not depressed – to a chemical with a faint odour, gradually increasing the concentration of the chemical until the volunteers could smell it. At the same time, the volunteers' brains were scanned to measure the volume of their olfactory bulbs. People without depression were able to detect the chemical at significantly lower levels than those with depression. The volunteers with depression were less responsive to the odour, and they were also found to have smaller olfactory bulbs in their brains, on average 15 per cent smaller. The more severe the depression, the smaller was the olfactory bulb. After successful treatment for depression, these deficiencies were reversed, and the ability to process smells and the size of the olfactory bulb returned to normal.

We do not yet know the reason why smell is affected by such mental conditions. One idea is that depression reduces the amount of a chemical in the brain that promotes the growth of new neurones – the so-called brain-derived neurotropic factor. Without this stimulus to grow new neurones, the circuitry dealing with smells, which normally is unusually vigorous in growing fresh neurones, shrivels. Another idea is that the amygdala, the emotional centre of the brain, is affected by depression, and that the close connections between the amygdala and the olfactory bulb mean that the brain is less responsive to smells.

The emotion that many people believe to be intimately connected to smell is love, or the more crude emotions of lust and erotic attraction. There is a billion-pound business founded on the hope that the smell of perfumes will make us attractive. There is a huge interest in the possibility that we communicate love and attractiveness by our natural smells as well as by our perfumes, and a prevalent fear that we may deter love by the off-putting nature of our natural smells. And there are pheromones.

A pheromone is a chemical given off by a creature which acts on the behaviour or physiology of other creatures of its species. There are alarm pheromones, trail pheromones, sex pheromones and many others. Ever since pheromones were discovered to be a powerful means of communication for many insects and animals, the search for human pheromones, and particularly sex pheromones, has been on. It is a compelling idea: that, without knowing it, we may be secretly communicating to the opposite sex and they with us; that we, like moths, are transmitting and receiving secret molecular messages. We love a conspiracy theory, and here is a theory that is about sex and power, and one that might just be real.

Many animals have an auxiliary smell system, alongside but separate from their main sense of smell. They have an organ – called a vomer – which detects pheromones, and nerves which communicate from the vomer to the brain. You can see this vomer in action in the domestic cat: when it makes a face by wrinkling its nose and pulling back its lips, it is directing scent to its vomer. This special face is called a 'flehmen'. You may see it too with a snake which flicks its tongue out and back, placing scents on its vomer. When you look for it, you will see many animals working their auxiliary smell system, flaring their nostrils and flicking their tongues out and back.

To the great satisfaction of conspiracy theorists, vomer-type organs were found on humans – at least on human foetuses. The organs go under the name of Jacobson's organ, and take the form of two tiny dead-end channels with their openings at the base of the nostrils. They are very small and can only be seen with a microscope. As we grow and develop, however, these channels atrophy, and in the majority of adults they disappear. In those few people who keep a recognisable vomer-type organ into adult life, the organ has no nerves and is not connected to the brain. When one looks inside the brain for a structure which would deal with information from Jacobson's organ, it cannot be found. It looks likely then that Jacobson's organ is a leftover from a much earlier stage of evolution, much like the reflex of newborn babies to clutch with their feet. Like humans, our closest ape relatives – gorillas, chimps and orang-utans – lack a working vomer-type organ, which suggests that evolution has been slowly editing out the auxiliary smell sense in our species for some time.

Such thumpingly negative evidence is far from the end of the story. Even without proof that we are equipped to detect pheromones, many people including some scientists continue to hanker after the hope of finding that we are secretly controlled by pheromones, perhaps through some hidden mechanism that has not yet been discovered. There are those who place their hope in other tiny olfactory organs, to be found with the help of microscopes up our noses. Some contenders are the Grueneberg ganglion, and the septal organ of Masera, or again it is possible that our main organ of smell has receptors for pheromones, though none have been identified so far. There are also tantalising fragments of evidence that, even if we cannot find the mechanism for detecting human pheromones, we nevertheless respond to hidden messages which might derive from pheromones.

One such fragment of evidence came from the University of Chicago in the 1970s, where a research project seemed to show that women who live in close proximity end up synchronising their menstrual cycles, through the agency of chemicals in their sweat. The McClintock effect, named after the main researcher at the University of Chicago, appeared to show that women change the rhythm of their menstrual cycle when exposed to the sweat of other women: sweat produced by women before they ovulate seemed to slow down the cycle for other women, while sweat produced after ovulation seemed to speed it up. The next cycle for all the women concerned would be more closely synchronised, until in the end they would be exactly synchronised. It looked like dramatic evidence of a hidden chemical effect that might involve pheromones and Jacobson's organ. Scientists however have an annoying habit of trying to replicate evidence from experiments, and attempts to replicate the McClintock effect have not had the hoped-for result. The jury is out on whether human beings can detect and respond to pheromones; as time passes, the jury is losing interest.

There may be some faint evidence in favour of pheromones being at work in newborn babies feeding from their mothers' breast. It has been found that rabbits produce a lactating pheromone, which is followed by the baby rabbit to find its mother's milk. The same researcher who worked on rabbits turned his attention to human babies.[19] He found that newborn human babies are also attracted to scents originating in the nipple area. These are very complex smells, containing secretions from the skin of the breast as well as smells coming from the mother's milk, not to mention the scents in any creams or lotions applied by the mother. So far none of the molecules examined can qualify as a pheromone. But if a human pheromone does exist, this is probably the area where we can expect such a discovery. If it were to be firmly established, it might prove consistent with the presence of Jacobson's organ in tiny babies, and with the organ atrophying as the baby develops. It would not however be a discovery of enormous importance, since it is no surprise to find that smell plays a role in bonding babies with their mothers; if we were to ask any new parent what smells they most enjoy, the answer would be likely to be the smell of newborn baby.

Whether human pheromones exist or not, conventional smells, coming to us through our conventional organ of smell, have a role to play in love. It is an unexpected role. In sniffing out a suitable partner, we are influenced by an olfactory factor of which we are unlikely to be aware. It has been known for some time that the immune system of our bodies can identify alien things such as a virus or a surgically implanted kidney, and that it can muster antibodies to fight them. A segment of our DNA – our genetic code – contains the instructions for these disease-detecting structures in our immune system. This segment is made up of the MHC genes (the major histo-compatibility complex) which carry the instructions to make proteins found on the surface of cells that help the immune system recognise foreign substances. These genes are passed on to our children, not, as with many genetic inheritances from one parent only, but from both parents. So if a child inherits a version of an MHC gene which gives resistance to Disease A from its mother, and a version of an MHC gene which gives resistance to Disease B from its father, it will be able to resist both diseases. There is therefore a great advantage to a child if its mother has chosen to mate with a man who has a dissimilar set of MHC genes. The mother does well to choose a man with a different immunotype.

The point of this genetic digression is that these immunity genes can be detected by smell. It was found from laboratory mice that a female mouse who is offered a choice of two mates will invariably choose the mate whose MHC genes are most dissimilar to her own. The female mouse evaluates the immunotype of the male mice by sniffing their urine, in which the different scents of each MHC gene can be detected. By choosing the mate with a different immune profile, the clever mouse secures for her offspring an advantage by ensuring that they will have resistance to a wider range of diseases. It is not, we can presume, a conscious choice in which the mouse is identifying each of the MHC genes by smell, but an unconscious choice driven by her preference for one kind of male scent over another.

That all this applies to human beings is a surprise. After all, we do not actively sniff each other's urine. Nor do we (nor the mice, when given the chance) choose our mates in laboratory experiments. But women are sensitive to the smell of men, and their smell preferences do line up with the immunity profile of the men in the same way as do the smell preferences of female mice. Women prefer the smell of men whose immunotype varies from their own. They tend to find that the smell of men with a similar immunotype to their own is offputting, and, when asked, they tend to explain that the smell is like that of their father or brother.[20]

Smell, then, plays a role in love, not through the agency of mysterious pheromones, but through the nose in the usual way. It plays a role for women more than men; and indeed women are generally more sensitive sniffers than men. It plays an entirely natural role, and the effect of perfumes may confuse, by masking the natural smells that guide the natural choice of mate. It may be that wearing perfume, instead of saying 'I am attractive', is really saying 'I am already mated with someone who values me enough to mark me with this expensive scent.'

Finally, we should touch on a completely different emotion which seems to have a special connection with the sense of smell – suspicion. In many languages, to 'smell' is to suspect, to 'sniff around' is to investigate distrustfully, to 'be suspect' is to give off a smell. In English, we say that something suspicious smells 'fishy' or we 'smell a rat', though in other languages the connection may be to the smell of rotting vegetables. Researchers at the University of Michigan[21] have investigated the suspicious nature of fishy smells, and they have found, yet again, that the connection between smell and emotion works both ways. Fishy smells make us suspicious; and being suspicious makes us more sensitive to fishy smells. So they tested the ability of subjects to spot misleading information and to distrust it. A slight whiff of fish oil in the air, they found, made it more likely that people would scrutinise dodgy information with suspicion, and less likely that they would accept it unthinkingly. They also created the opposite scenario – by getting a researcher to act suspiciously during a smell test, they found that people whose suspicion was aroused were more likely to discern fishy smells, and better able to label them correctly. An atmosphere of suspicion sharpened their response only to fishy smells, not to other smells. Fishiness really is an attitude as well as a smell.

CHAPTER 8

A False Scent

'Where two or three thousand words are insufficient for what we see … there are no more than two words and perhaps one-half for what we smell. The human nose is practically non-existent. The greatest poets in the world have smelt nothing but roses on the one hand, and dung on the other …'

Virginia Woolf – *Flush*

It is a common enough belief that the human sense of smell is completely impoverished when compared to that of sniffer dogs and bloodhounds, of rats and mice, of sharks and other creatures that hunt by smell. It is a common belief, but it is misguided.

Almost every living creature has a sense of smell, and the sense has functions that vary enormously from creature to creature. We tend to think first of animals that hunt by scent and can follow scent trails, and to regard them as having very superior powers of smell. The sense is well-adapted to hunting in animals that hunt like this, certainly. But it is well-adapted to defence in prey animals, to navigating in migratory creatures, to social organisation in social creatures, and so on. When we pause to consider what functions the sense of smell serves for us humans, it is quite possible that we will find that those functions are different to those of a sniffer dog or a bloodhound or a bee, and that, rather than being a third-rate version of the senses of dogs or bees, our sense of smell is eminently fit for our purposes.

Even the tiniest creatures have a response to smell. Volvox is a tiny primitive organism that forms colonies of algae in puddles and ponds. It is credited with being the most primitive creature that goes in for sexual reproduction: the simplest creature that does not reproduce only by splitting itself into clones. As if being the most primitive creature that goes in for sex were not enough, Volvox is the most primitive creature to suffer natural death, rather than to ensure survival by cloning itself. Primitive as it is, with no organs worth speaking of, Volvox responds to chemical signals – to smells – in its puddles and ponds.

Only one class of creature appears to have no sense of smell – dolphins and porpoises, the 'toothed whales'. They have no smelling apparatus, neither in their noses nor their brains. They rely on other, brilliant senses, notably echolocation. Dolphins live in water but breathe air, and it seems that in the process of evolution, which equipped them with the ability to make use of the two environments, their sense of smell was edited out. As the antecedents of dolphins evolved into marine animals, and their breathing hole migrated from the front of their heads to the top of their heads, the connection between breathing and smelling was lost. They no longer needed to smell the air, since they live underwater, and they had no need to smell the water since they could listen to it so effectively.

Amongst the many creatures with excellent smelling apparatus, it is striking how many varied functions are served by their senses of smell. One can find astounding examples of sensitivity to smells for one reason or another. But one cannot find a creature which carries away all the prizes for excellence.

It is certainly impressive that there are hunting animals that can follow scent trails that are days old, or can sniff out substances that are undetectable by other creatures. But there are also impressive applications of the sense of smell to homing, to social organisation, to attracting prey, to deterring predators, to bonding, to managing a territory, to alerting others to danger, to discriminating edible food, to establishing status in a group. For many creatures, smell is not just for sensing but also for communicating. So let's pick out some creatures which make striking use of each style of smelling. In that way we will see how many purposes are served by the sense of smell, and how many forms of excellence there can be.

Dogs are often thought of as the most talented hunters by scent. But polar bears can track scent over longer distances, hunting as they do in a frozen desert where smells travel further than sights or sounds. A polar bear in Alaska has been tracked following a wind-borne scent to a seal's breathing hole. It travelled 40 miles in a straight line over the tops of pressure ridges in the ice and over open leads in the ice to reach the seal's hole. Its usual hunting method is to locate a seal's hole and then to lie hidden close by until the seal surfaces to breathe. Then, as soon as it smells the seal's breath, it strikes with its massive forepaw. Another hunting method is for the polar bear to sniff out a seal's hibernation den beneath the ice and use its power and weight to punch through the ice. The task for a polar bear's nose is even more demanding when the prey is in its hibernation den, as the bear has to sniff out the prey under a metre or more of ice and from a considerable distance away.

Another creature that hunts by smell, but on quite a different scale and using quite different techniques to those of a polar bear, deserves a mention. This is the innocent kiwi of

New Zealand. It sniffs out underground worms by night. One curiosity is that the kiwi's olfactory nerves are towards the tip of its long beak, rather than at the back, so that it can probe for smells within the soil. This hunting style gives rise to a second curiosity – the kiwi's habit of snuffling like an old gentleman with nasal problems. It snuffles to clear its beak of earth so that it does not become clogged as the bird probes for worms. A simple experiment has demonstrated the accuracy of the kiwi's hunting technique. A series of deep flower pots were filled with earth, some containing worms at the bottom, some empty of worms. The pots were covered with plastic film. Overnight the kiwi visited, and its activities were revealed by perforations in the plastic film covering the pots. Only pots containing worms had been perforated; and there were only as many holes as there were worms. The kiwi had unerringly found its prey by smell.

Smell is used by predators to find prey; and frequently the prey can fight back with smells. Skunks are probably the most notorious practitioners of self-defence by stink. Their scientific name – *mephitis mephitis* – is taken from the Latin for a stink. They spray from an anal gland which can squirt accurately at a range of two metres: some species of skunk do a handstand to deliver their spray more effectively. The stink is sulphurous, so strong it can cause retching, and it is also an irritant which can cause temporary blindness. It is widely believed that the only way to get the stink out of a pet that has been sprayed by a skunk is to bathe the pet in tomato juice. Charles Darwin, in his travels on the HMS *Beagle*, met South American skunks called zorillos:

> We saw also a couple of Zorillos or skunks – odious animals, which are far from uncommon … Conscious

> of its power, it roams by day upon the open plain, and fears neither dog nor man. If a dog is urged to the attack, its courage is instantly checked by a few drops of the fetid oil, which brings on violent sickness and running at the nose. Whatever is once polluted by it, it is for ever useless ... The smell can be perceived at a league distant; more than once, when entering the harbour of Monte Video, the wind being off shore, we perceived the odour on board the Beagle. Certain it is, that every animal most willingly makes way for the Zorillo.

Among the smelliest birds is the otherwise charming hoopoe, a pretty visitor in early summer to southern Europe from Africa. The young nestling hoopoe has a large preen gland which produces a stench like that of rotting meat. The gland comes into action shortly after the young bird emerges from its egg, and ceases to work as it fledges, so there is little doubt that the function of the smell is defensive. Smelling bad may help birds in other ways, for instance by keeping their feathers and skin free of parasites. A study found seventeen anti-microbial compounds in the wood hoopoe's foul excretions, serving to protect it from lice, bacteria, moulds and fungi.

A different form of smell defence is demonstrated by minnows. When a minnow is threatened or injured, it releases a chemical signal – an alarm pheromone – which serves to warn other minnows of danger. Many fish have now been found to send out chemical signals of this kind. They elicit a variety of responses: some fish skitter or dart while others freeze. Some fish huddle together in shoals, while others scatter. Some species respond to alarm pheromones by swimming nose down to stir up mud, while others rise to the surface or even jump out of the water.

Many creatures use smells to attract sexual partners. But the prize surely belongs to the emperor moth. According to German experiments in 1961, the male emperor moth can detect the sexual attractant of the virgin female at the extraordinary range of eleven kilometres. Only the minutest amount is released into the air at any one time. The male moths sense the odour of the female with their feathery antennae, which are so sensitive that they can detect a single molecule at a time. They can also assess its strength and, as it increases, the moth is able to navigate towards the female.

Attractant scents – 'sex pheromones' – can also be used by predators as a tool to kill moths. The bolas spider synthesises a chemical that mimics the female sex pheromone of the cutworm moth, and broadcasts that chemical in the evening when the male moths fly, luring them in to be caught and eaten. After midnight, the same spider switches over to dispense the pheromone of a different moth species – the Smoky Tetanolita – which is active only later in the night. This small genius of a spider is named because its technique for catching prey is like the bolas used by South American gauchos in place of a lasso to catch cattle. The gauchos' bolas is a weapon made of a rope with two or three balls attached, which are whirled around and hurled at the legs of cattle, to entangle them and bring them down. So it is with the bolas spider, which, instead of building the conventional web, dangles from one leg a length of silk that has a blob of sticky glue at one end. When the moth is within range, the spider swings its 'bola' at the prey to entangle it.

A quite different function of smell is in defining territory, whether in helping an animal locate its home territory or in allowing an animal to warn others off its own territory. The prize for homing by smell should probably go to salmon. That salmon swim up streams from the ocean to spawn in fresh water was known for centuries. But it was not known that they were returning to where they were born, nor that the spawning migration might be thousands of miles long, nor was it known how they found their way. Experiments in the USA in the 1950s[22] established that rivers have unique odours, and that salmon can distinguish between them. Salmon are imprinted with the smell of their native stream early in life, and retain the memory through to the end of their lives. The experiments established too that when a young salmon is raised in water that has been given a synthetic scent, it will navigate its way back to a stream which has been dosed with the same synthetic scent. There were also experiments which established that, when the researchers blocked the smell organs of salmon, they were unable to find their way to their native streams. Between the time that a young salmon learns the scent of its native stream and the time they trace the scent and find their original spawning ground, the salmon will have travelled thousands of miles across the ocean, using senses other than smell to navigate. Like sailors of old, they judge their position by assessing the height of the sun, the length of the day, the salinity of the water and the magnetic field of the earth. But once they are within sniffing distance of fresh water, the salmon follow their noses.

A humbler demonstration of homing by scent is given by the limpet. The limpet is well known for clinging to rocks and for unshakably withstanding the force of the waves and tides which batter their chosen place. It is less well known that limpets release their grip on the rock at low tide, and trundle off to feed, equipped with little more than a set of 1,900 teeth and a smell organ called an osphradium. They feed up to three metres from home, and then as the tide turns they sniff their way back to their original emplacement. It has to be the exact spot, with the exact orientation, so that the limpet can make a perfect airtight fit to prevent the waves from sucking it off the rock. The limpet manages this little feat of navigation by smell.

Smell is important to many hunting animals in marking their territory and warning others off their patch. Hyenas provide a prime example of the use of scent in marking territory, but also in establishing the identity of the pack. Hyenas roam large territories in packs, and use smell to guide their way. Pouches in their rear ends produce a foul-smelling paste called 'hyena butter' which they rub onto vegetation and onto each other. The smell does not just mark territory; it is also used by other hyenas in the pack to keep up to date on its membership. The hyenas gain a clan scent by rubbing their bottoms against vegetation that has already had a scent deposited by another pack member, mixing their own scent with that of others in the clan. As clan membership does not stay the same, the hyenas must regularly anoint themselves with the scent of others in the clan to retain their membership. An analysis of the chemical composition of hyena butter showed that not only do individual hyenas have a unique scent, but that there are components in the smell that are shared within the clan but not by other packs.

Scent also communicates status. Ring-tailed lemurs live in large social groups. During the breeding season, the males compete fiercely for mates. But rather than battling each other using their sharp teeth and claws which would put them at risk of serious injury, they have a stink-off. Male ring-tailed lemurs use scent glands on their wrists and shoulders to send a warning to rivals. The wrist gland produces a short-lived smell, while the shoulder gland produces a brown paste with a much longer-lasting smell. The duel begins when two lemurs face off and comb the musk from their wrist glands into their tails. They then flick their tails at their rivals, wafting their smells towards each other. The stink-off can last an hour, and only ends when one lemur backs down. Stags do something similar, when they thrash the vegetation with their antlers in a mock fight with a rival, and in the process spread their individual scent around the area.

Nowhere is scent more important than in coordinating the societies of social insects. In honey bees, for instance, the absence of a specific scent causes the workers to build wax cells to hold queen larvae; another scent emitted by a queen bee in flight attracts drones to mate; and yet another of her scents depresses the ovary development of her workers so that she ends up as the sole egg layer in the hive. Honey bees also use scent for homing. When a swarm of bees flies, guided by its scouts, to a new home site in a cloud of perhaps 10,000 bees, most of them will not have been to the new site before and therefore have no way of knowing where its entrance is. So scouts perch at the entrance, lift the tips of their abdomens high into the air, open glands which release a marker scent (which to us smells lemony), and beat their wings to broadcast an odour plume to guide the others to them. There may not be much of a breeze to make a scent trail for the swarm to follow, but the bees that plant themselves at the entrance to the new home create one by fanning their wings. They may also align themselves into chains, and so create a directional flow of air. The other bees follow the scent trail and are guided home. Until the bees have thoroughly learned where home is and can navigate to it, some bees will remain on duty to mark the entrance with that lemony scent.

One kind of social insect, the army ant of South America, uses smell to create a weapon of mass destruction. Their capacity to mark a trail and to mobilise an army to march along the scent trail in a column is close to terrifying. Army ants have been known to devour large animals that fail to get out of the way of a colony on the move – anything from poisonous snakes to tethered horses. There is a report (or more likely myth, though it is difficult to be entirely sure) of a column of army ants a mile long and half as wide marching on the town of Goiandira in central Brazil in December 1973. It is reported that the column devoured several people before being driven back into the jungle by a team armed with flame throwers.

What emerges from exploring the sense of smell in other creatures is that it is a sense with many possible uses. It is for detecting things, as we know well, but it is also for communicating things, for interpreting the complexities of social relationships, for attracting and deterring, for eliciting behaviour in others. The special characteristics of this sense, which often to us seem deficiencies in comparison with vision or hearing, make it particularly valuable. Unlike sight, smell does not travel in straight lines, so it is valuable in environments where sight does not serve well, such as thick jungle or water or in a mass of insects. Unlike sight or sound, which can be interpreted only within a narrow range of visible light or audible sound, smells come in a myriad of varieties. So an animal can tune in to a very particular scent that may mean nothing to other species; a single molecule is enough to convey a particular message to a particular species. It is a very specific sense. Unlike sights and sounds, smells linger, so they can be deposited as a marker while the creature goes about its other activities. Unlike sight and sound, smell does not require an elaborate sense organ – at its simplest it can be a few sensory cells linked to a central nervous system.

What then of the sense of smell in us humans? We cannot claim prizes for following scent trails, nor for navigating by smell, nor for sniffing out a sexual partner at eleven kilometres. Most of us assume therefore that we have an inferior sense of smell, a fourth-rate, low-budget, close-to-redundant sense. Yet our sense of smell serves our particular needs well. In some respects, it performs feats of which no other creature on the planet is capable.

Four purposes seem to be met by our human sense of smell. First it serves to warn us of danger, with a particular sensitivity to very small concentrations of smoke, as you might expect for an animal which, unlike any other, plays with fire. Second it serves to tell us about home ground, with a particular responsiveness to familiar and unfamiliar smells, as you might expect for a very territorial, home-making animal. Third, it serves as a form of social communication, as you might expect for a social animal. It bonds us to our kin, and it tells us about our fellow humans, their gender, status, aggression. Fourth, our sense of smell gives us very fine discrimination particularly of the smells of food and drink, as you might expect for the only animal on the planet which cooks its food.

What makes our sense of smell special is that it is used in combination with our ability to make things. It is a fair bet that humans are the only creatures that can fabricate smells synthetically. It is a fair bet that only humans can distinguish China tea from Indian tea from Ceylon tea, or Syrah wine from Merlot wine from Rioja, or baked potatoes from roasted or fried or boiled potatoes. I know of no other creatures that scent their homes for the pleasure of it rather than for defence or recognition, or that make use of smoke for pleasure or religious worship, or that use their sense of smell in the flavouring of their food and drink.

In many ways, we may have given ourselves an inferiority complex by our close association with 'man's best friend'. Several breeds of dog have an exceptional interest in smells and are equipped with a muzzle that is longer and more complicated than a human nose and that therefore gives them more control over incoming smells than we humans have. Furthermore, they have been bred and trained by humans to make the most of their noses. It is not surprising then that we should be impressed by the sniffing abilities of an animal that we have deliberately engaged as a sniffer.

The superiority of dogs in following a scent trail is not diminished by the discovery that human beings too can follow a scent trail. In one of the weirder experiments in the field of olfaction, humans were set to follow a chocolate-scented trail. At the University of Berkeley California in 2006,[23] 32 students were dressed in coveralls, a blindfold, earplugs and gloves, to exclude all sensory clues except smell. They then got down on all fours to follow a 33-foot trail of chocolate through the grass. Most of the subjects were able to follow the chocolate scent to the end of the trail within three attempts, zigzagging along the trail in exactly the way that tracking dogs follow a scent. Those volunteers who gave time to improving their performance became nearly as fast and unerring as a tracker dog.

Nick Carter was a bloodhound. Born in 1899, he was named after a popular pulp fiction detective of his day. Nick Carter was trained in the first place as a tracker of escaped convicts in Kentucky. He became famous for his ability to follow cold trails – those that were days old – and once followed a trail that was twelve days old. On another occasion, he followed a trail for 58 miles. Crowds would gather when it was known that Nick Carter would be set to a trail. Indeed, it became a problem for Nick's handler to keep the crowds from interfering with the bloodhound at the point when he first had to pick up a trail.

Nick Carter's greatest hour came with a case of suspected arson. A barn had been burnt down four days before. The charred ruins, stinking of old ashes and cinders, and much trampled around by people trying to fight the fire, should have offered no clear lead to the bloodhound. Nose to ground, the dog circled the barn. 'Then suddenly he lifted his tail and was off.' He led his handler to a farmhouse a mile off. The culprit was inside, and admitted his crime. By the time Nick Carter's legs became too old for trailing men, he had made 650 finds.

The snouts of scenting dogs have several features that give them an advantage in finding and following a scent. One obvious feature is the moist black tip of a dog's nose, with nostrils that can be twitched independently of each other to locate a scent. The black muzzle supplies mucus which helps in packaging smell-bearing molecules on their way to the dog's olfactory nerves. A dog that is running short of mucus will lick its muzzle to freshen it up. More significant is the internal structure of the dog's nose, which contains a curl of thin bones, shaped much like a brandy snap. These bony convolutions, called the ethmo-turbinal and maxillo-turbinal bones, form a kind of cartridge through which incoming air is filtered and warmed, preparing it for the nasal receptor nerves. At the upper end of this cartridge of bone is a crossways bone which forms a shelf to trap smelly molecules, effectively giving the dog time for a good think about what it is smelling. Human noses have a much simpler structure, one which is less well-adapted to sniffing incoming air, but which nevertheless has different advantages.

A feeling of being inferior to dogs in the matter of scenting is at the heart of a sparkling but misleading book by Virginia Woolf. Her book *Flush* is the story of the love affair of the poets Robert and Elizabeth Barrett Browning seen from the viewpoint of Elizabeth's pet spaniel, called Flush. Flush experiences the world through his nose. At first he is a country puppy, ill-treated until Miss Barrett acquires him and takes him into her protected indoor life in Wimpole Street in London. She is treated by her family as an invalid, unnecessarily so, and Flush therefore gets little opportunity to experience the odours of London streets. But when Robert and Elizabeth fall in love, and elope to Florence, then Flush is free to wander Italian streets and savour Italian smells.

> Flush wandered off into the streets of Florence to enjoy the rapture of smell. He threaded his path through the main streets and back streets, through squares and alleys, by smell. He nosed his way from smell to smell; the rough, the smooth, the dark, the golden. He went in and out, up and down ... with his nose in the air vibrating with the aroma. He slept in this hot patch of sun – how the sun made the stone reek! He sought that tunnel of shade – how acid shade made the stone smell! He devoured whole bunches of ripe grapes largely because of their purple smell; he chewed and spat out whatever tough relic of goat or macaroni the Italian housewife had thrown from the balcony – goat

> and macaroni were raucous smells, crimson smells. He followed the swooning sweetness of incense into the violet intricacies of dark cathedrals.

Virginia Woolf goes on to denigrate the human sense of smell by comparison with that of dogs, and in the process recites some common misconceptions about human abilities. 'Where two or three thousand words are insufficient for what we see … there are no more than two words and perhaps one-half for what we smell', she writes. Now which two and one-half words would those be? Is it smell and scent? Or is it perfume, reek, a whiff, pungent, sniff, mouldy, aroma, to nose, stench, stink, taint, odour, waft, bouquet, or is it fuming, peppery, acrid, redolent, floral, resinous, fragrant, putrid, heady, earthy, aromatic, rancid, tangy, piquant, pongy, fetid, gamey, sulphurous, incense, goaty, ammoniac, astringent, catty, pissy, foxy, mephitic, musky, musty …?

That barely scratches the surface, because there is an array of flavour-laden words that work to describe smells. They may seem to belong to the world of taste, but taste, smell and flavour are so closely wedded that they cannot be separated. So smells can be minty, smokey, sweet, bitter, sharp, sour, cheesy, fishy, spicy, citrussy, treacly, zesty, medicinal, oaky, herbal, honeyed, fruity, burnt, nutty … and so on. We still have not mentioned the words for smells that are tactile. But smells can be warm, cool, creamy, gummy, waxy, powdery, rounded, clinging, cloying, dry, damp, soggy, fresh.

Virginia Woolf goes on: 'The human nose is practically non-existent. The greatest poets in the world have smelt nothing but roses on the one hand and dung on the other.' Here she suggests another falsehood, that if the greatest poets in the world (Shakespeare, I presume, since he wrote about a rose by any other name and of stinking dunghills) have a limited perception of smells, then so, too, must the rest of us have a limited perception of smells. This assertion purveys the false belief that, between the extremes of floral sweetness and dungy foulness, we are not able to distinguish smells. In reality, the human nose is capable of very fine discrimination. I have only to pick out some herbs and spices from the kitchen cupboard, or visit a hardware shop, or walk into a school smelling of polish and children to realise that my nose can distinguish many shades of aroma, and there is a wealth of smells in between roses and dung. It may even be, we do not know for sure, that the human sense of smell is capable of finer discrimination than that of many other animals.

A view that animals have wonderful powers of smell, and that humans have inadequate powers, is wide of the mark. Humans are creatures with senses adapted to our way of life, one of a myriad of species with varied ways of life and variously adapted senses. To look for extreme examples of olfactory power is to ignore the many species which possess powers that are admirable but not remarkable.

CHAPTER 9

Two Halves of Bitter and Four Packets of Crisps

> 'Smell and taste are in fact but a single composite sense, whose laboratory is the mouth and its chimney the nose.'
>
> Brillat-Savarin

I love the authentic soggy smell of an English pub. It is a quintessentially English smell, homely and at the same time slightly illicit. It is a typically Irish smell, too, and James Joyce caught it perfectly in *Ulysses* when his hero Leopold Bloom saunters past a Dublin pub – 'the flabby gush of porter … whiffs of ginger, teadust, biscuitmush'. I wanted to experience that flabby gush of beer and at the same time to explore the obscure world of retronasal olfaction – the hidden talent for smell that humans have (and dogs do not).

I narrowed down the choice of pub to the elegant Falkland Arms, with wine list and function room, and the squat Duke of Cambridge down the road, with a giant England flag draped over the window, red cross on dirty white, and a couple of smokers just outside the front door. There was no question which to choose: for a genuine, old-English, no-frills pub, untouched by New Zealand wine and Arabica coffee and gastronomic cooking and leather settees, it had to be the seedy Duke of Cambridge. I entered at a dull time of day, late morning, to find two delivery drivers at the bar with the elderly landlady and a couple of old-age pensioners silent on the window seat. A gambling machine winked against a wall; a TV with the sound turned down gave us the midday news; the ceiling and walls had a crusted wallpaper varnished with the nicotine of ages. For the sense of smell the pub offered, first, the sour tar of old cigarettes, as I entered past the zone of smoking just outside the door. Second, it offered a sharp unperfumed disinfectant, swabbed over the bar and floor as part of the early morning cleaning, mingling with the flabby gush of beer spilt on the bar and on the floor, evening after evening, year after year.

I ordered a half of bitter from a pull that bore the emblem 'Malty and Tropical'. I could imagine the sharp young marketing person at the brewery who had brought that touch of modernity to the unvarying traditional business of brewing bitter. She (or he) was taxing the language to its limits, for there is nothing tropical about the watery mush of barley, hops and yeast that makes an ordinary English bitter. Tropical it was not, but it was malty and it met in full my need for comfortable familiarity.

I sat on the window seat next to Brian, an elderly man whose legs were playing up. He needed a Zimmer frame, and grumbled about his legs, his mobility and his doctors.

On his tottering return from a smoke, I offered Brian a beer, and engaged him in conversation on the sense of smell. 'Not what it was, my sense of smell,' he grumbled. Machine oil (he had been an engineer), cigar smoke, lakeside reeds (he loved fishing) and bacon were his favourite smells. No, he had not given much thought to analysing smells, and he meant no offence by saying so but he considered my project bloody weird.

By stages we moved through the difficulties and pleasures of Brian's sense of smell, as I artfully steered the conversation towards my goal of the day. This concerned the role that our noses play in the flavour of food and drink.

I wanted to explore what goes on when we taste things, whether it is our taste buds that do the work, or whether other senses are in play. The literature tells us that flavour – the gustatory experience – is a subtle concoction of the simpler tastes discerned by the tongue with the more complex smells of the food discerned by the nose. Beyond these two senses, the feel of food, the expectations of flavour created by the look of food, and the sound of crisp or liquid chewing all play a part.

So Brian was the subject of a simple experiment. I bought four packets of different flavoured potato crisps at the bar – plain salted, prawn, smokey bacon, and salt-and-vinegar – and put one crisp from each packet in front of Brian. I asked him to hold his nose, and tell me which was which. 'No difference,' he said, '– all plain, just salt.' He then released his nose, and this time, found the flavour. He reran the experiment with crisps of each flavour, and found how little was perceived by the taste buds on the tongue, and how much was perceived once the nose was in play. He became excited, and enlisted his elderly neighbour in the experiment. Then the two lorry drivers at the bar joined in, and the landlady, and we all held our noses, crunching our crisps and sipping our drinks, releasing our noses, exclaiming over the difference.

I felt a school-masterly moment coming over me. I explained that what is happening when we release our noses and receive the full force of our synthetically-flavoured crisps is called 'retronasal smelling'. It is the reverse of normal smelling. Normal smelling is done on an in-breath. Retronasal smelling, backwards smelling, is done on an out-breath. The molecules of food and drink in our mouths, warmed up by our body temperature and stirred up by chewing and saliva, fill the airways in the back of our mouths which communicate with our noses via the nasopharynx – the large passage that joins the back of the mouth with the nose. The strong movements of chewing and swallowing pump the molecules into our noses, and breathing out brings the molecules into contact with the receptors of smell. This is the moment when our sense of smell contributes its sophisticated powers of discrimination to the business of sensing flavour. By holding our noses, we prevent the air from circulating in our noses, and so prevent retronasal smelling.

It is astonishing, I continued, warming to my theme, that most of us chew away meal after meal, comfortably convinced that the flavour of every mouthful is located in our mouths and is produced by our taste buds. This is a daily, nay, a thrice-daily illusion. The experience is a referred sensation, referred to the mouth but more justly ascribed to our noses. It is a referred sensation like the experience of vertigo in the base of our spines, or an itch that cannot be scratched, or the experience of phantom limbs. Taste is the junior partner of flavour, and the taste buds are the rather clottish clerks in the front office, while the real work is done by the subtle senior partners in the nose.

Whatever the vaunted talents of dogs, rats, polar bears and moths (I continued), there is no evidence that any animal other than us human beings makes use of retronasal smelling. We are in this one respect at least superior to other animals despite their amazing abilities in tracking scents and in interpreting each others' urine. The animals that can follow scents have noses that are better equipped than ours for sniffing the ground, but we have noses better equipped than theirs for retronasal smelling. In human noses there is a relatively short distance from the back of the mouth to the sensory nerves in the nose, and a relatively large opening from the back of the mouth to the back of the nose. This allows the odours released by chewing food to move easily from the back of the mouth to the nose. Humans are also able to roll their tongues, unlike other animals apparently, and so we are able to manipulate our food with our tongues and to extract flavour from it while chewing. We also use our tongues while swallowing to create a temporary airlock in the mouth, and so to pump the flavour-laden air around the back of our noses. Other animals gobble their food and drink; we alone savour it.

I was in full flow on the intricacies of flavour when Brian's wife came into the pub. She had been shopping while Brian was parked in the pub with his half pint. Amongst her shopping she had a bag of fresh cherries. When Brian explained to her the nature of our experiment with potato crisps, she offered her bag of cherries to us. We each took a cherry. We solemnly held our noses and bit. We looked at each other baffled – the flavour of the cherry without retronasal smelling was as close as dammit to the flavour of a cherry chewed normally. I felt abashed, and made my exit soon after.

I hastened away, determined to get more fully to grips with the science of retronasal smelling. The popular literature tells us that 80 per cent or 90 per cent of flavour comes from the sense of smell. Clearly, my botched experiment at the Duke of Cambridge had shown that such a number as 80 per cent can only be guessed at, and cannot be measured precisely. If the contribution of smell to flavour varies from food to food, the proportion of flavour that can be attributed to smell will vary from person to person, depending on what foods they habitually eat. A habitual cherry eater will receive more of his flavour from his tongue than a committed eater of synthetically-flavoured crisps.

Back home I considered how to shine light on the contribution of smell to flavour. I decided to devote a day to eating without retronasal smelling, in order to work out what contribution the nose makes to flavour. With each mouthful of food I took, I first held my nose to find out how it tasted without the sense of smell, and then released my nose to assess what elements of flavour were exclusive to the nose. By the end of the day my nose was swollen with being pinched, but my understanding of flavour was transformed.

I started at breakfast. My customary cup of sugary, instant coffee felt comfortingly syrupy in the mouth. But, without the sense of smell, it could have been a sweetish soup for all the flavour I could discern. Only when I released my nose, and allowed retronasal smelling its opportunity did I get the characteristic flavours of coffee – nutty, bitter, roasted. It was a worse case with toast; without the sense of smell you are deluded by the feel of toast in the mouth into believing that you taste it. The scratch of toast against the gums, the snarling face you make as you bite down, the satisfactory crunch as you chew, all these tactile sensations fool you into having a sensation of flavour. It was only when I released my nose and received the full flavour of toast that I realised that I had received absolutely no sensation of flavour from my tongue; all of the nutty, malty, smokey quality of toast is recognised by the nose and not by the taste buds on the tongue. I added a good slab of butter to my toast, expecting that a good deal of the pleasure of butter would be in the feel of it, more than in the flavour. But again, I found that the flavour of butter has complexities that are not recognised by the tongue, but by the nose. It was only when I added marmalade to my toast and butter that my tongue perceived any real flavour, but even marmalade offered only a very partial experience without the sense of smell: it offered my tongue the sensation of sweetness and tartness, but curiously separate and incomplete. Only with the sense of smell did the flavour of marmalade come together into a complete blend – fruity-sweet, acid-sour, floral and fermented.

From this very meagre breakfast I had learned immediately that what we think of as taste is an illusion. Your brain constructs this illusion: that the sensation is located in the mouth, whereas in reality the sensation is constructed by several different organs in various parts of your face. It is illusory too, that this is just one sense – it is in fact a blending of several senses. Take away one or more of the main components of flavour, and your brain continues for a short time to fool itself into thinking that it perceives the full sensation. It is an inaccurate illusion: you guess what coffee or toast will taste like, and it comes as something of a surprise when your nose is finally allowed to give you the real flavour. In your imagination you have constructed a rough idea of the flavour, but it is missing many redolent components. It becomes obvious, when you play around with retronasal smelling, that most of the sensation of flavour is sensed by the nose, but even when that is obvious, you still have the illusion that the location of flavour is the mouth.

It is not a new discovery that flavour is constructed from many senses combining into a single perception, nor is it a new discovery that the location of flavour in the mouth is an illusion. These facts began to emerge in the 19th century. But serious scientific enquiry into how flavour is constructed is relatively new. The recent establishment of research centres working on flavour in a multi-disciplinary way in major universities like Oxford and Yale is a sign that the time has come for this subject to be unpicked. So too is the emergence of celebrity chefs who play deliberately on the multi-sensory aspects of flavour to surprise their customers or to enhance their experience. The illusion that flavour is experienced in the mouth is called 'oral referral', and is seen by psychologists as akin to the illusion pulled off by ventriloquists. We see a ventriloquist's dummy moving its mouth and we hear a voice, and so we assume that the moving mouth is producing the voice. In much the same way, we put food in our mouths and experience the taste, and so we assume that the mouth is producing the flavour. The latest research suggests that this is more than the ventriloquist illusion, because we delude ourselves not only about where the stimulus is located (mouth instead of nose) but also about the ensuing perceptual experience (taste instead of smell). Three factors may be at work: the fact that the food goes to the mouth first; our tendency to confuse smell and taste when the two are congruent; and our tendency to give smell less continuous attention. Researchers believe that the illusion of flavour is a phenomenon all on its own, a unique piece of multi-sensory perception.

I continued my regime of eating without retronasal smelling through the meals of the day and during a large number of experimental snacks. I ate vegetables cooked and raw; fruit raw, dried and fermented; meats and fish and fowl; dairy products; bread and biscuits and buns; spicy and herb-flavoured foods. Each mouthful I took first with my nose blocked, and then with my nose released; as I did so, I recorded the differences in flavour. Some fascinating general conclusions emerged.

Raw and cooked foods are completely different in the demands they make of the sense of smell. Raw fruit and vegetables give perhaps 50 per cent of their flavour to the taste buds, but there is still at least half of the story that requires the nose to complete. Cooked food gives hardly any flavour to the taste buds: all the flavours of cooking – the smokiness, the caramelised flavours, the blending of sweet and sour – are perceived by the nose and not by the tongue. Herbs and spices are exclusively for the nose to discern; the only sensation that they offer to the mouth is the sensation of heat, of the piquancy of ginger or pepper. Fermented foods and drinks, too, are meaningless without the sense of smell; wine and cheese and salami have almost no taste without retronasal smelling.

The sense of smell contributes the bulk of what we consider as flavour. And it performs four particular functions that are not performed by the sense of taste. First, the sense of smell catches nuances that elude the sense of taste. Taste gives us saltiness and sweetness and sourness, but it is the sense of smell that gives us the special saltiness of salami as opposed to the special saltiness of anchovies; the special sweetness of strawberries as opposed to the special sweetness of tomatoes; the special sourness of oranges as opposed to the special sourness of grapefruit. Second, the sense of smell alone gives us the flavours of cooked, cured and seasoned foods.

The third, and perhaps the most significant contribution of the sense of smell to flavour, is that it gives us a rounded sensation, whereas the sense of taste gives us an incomplete sensation of distinct elements of flavour. To make clear how the sense of smell rounds out the flavour of food, I held my nose while eating a fresh apple – one plucked from the tree, rather than an anaemic supermarket apple. It tasted juicy, sweet and tart, but it did not (until I released my nose) taste quite like an apple. The sweet and sour juiciness of an apple on the taste buds is much the same as the sweet and sour juiciness of a fresh tomato or a fresh orange on the taste buds. It is only when the nose is brought into play that the complete story becomes clear – that the apple tastes like an apple with a rounded blend of earthiness, of berry-like sharpness, of a sweetness that is reminiscent of pineapple, and of a touch of nut. There is another aspect of flavour that only the nose provides. This is the evolution of a flavour over time. As I chewed my apple, my sense of smell kept me informed of the changing nuances as the flesh of the apple and its skin released their elements.

Could it be that the human sense of smell is not, as many have asserted, a primitive sense that we are progressively losing as we evolve into more sophisticated creatures living in an increasingly visual world? Could it be, rather, that it is a highly-evolved sense that has adapted to millennia of specialised use? It seems an idea worth dwelling on – that we have evolved from being primates that ate a diet primarily of fruits and roots and needed a fairly simple response to flavour, into omnivorous creatures that eat food that needs cooking and seasoning, and that our response to flavour has become more complex accordingly. In service to our desire for flavour, the human sense of smell has evolved into a sophisticated, subtle and discriminating organ. Our noses may be simpler structures than those of a sniffer dog, but the capacity of our brains to get meaning and pleasure from flavour is probably much greater.

What emerges from this simple exercise in isolating some aspects of flavour is the wonderful complexity of the brain's work. It creates a single experience of flavour from five separate senses. Each of the contributing senses has its distinct array of nerves; each responds to its distinct set of sensations; each feeds its sensory information into different parts of the brain. But the brain blends this all together, and we perceive a single mouthful. Its tastes, its smells, its temperature, its feel, its intensity, its crunch or crispness, its sound, they combine into a single thing in the mouth.

The man who first wrote about the complexity of flavour, and at the same time founded the whole genre of gastronomic writing was Jean Anthelme Brillat-Savarin, a lawyer and politician who is remembered as an epicure and gastronome. In 1789, at the start of the French Revolution, he was sent as a deputy to the 'Estates General', the assembly of the country's ruling classes that soon became the French National Assembly. As the revolution turned bloody, he found that a price had been put on his head; he fled first to Switzerland and then to the United States where he lived for three years giving lessons in French and the violin. He returned to France in 1797 under the government of the Directory, and when Napoleon took power, Brillat-Savarin became a judge and a comfortable member of the Establishment. His spare time was devoted to the pleasures of the table, and at the end of his life he wrote *Physiologie du Goût*, a series of aphoristic reflections on taste.

His relevance to this book is that he fixed for the first time the importance of the sense of smell to flavour. 'Smell and taste are in fact but a single composite sense, whose laboratory is the mouth and its chimney the nose.' He reports meeting a man in Amsterdam, a Dutch sailor who had been captured by pirates from the North African – Barbary – coast and had been taken into captivity as a slave. The sailor's tongue had been cut out at the root by Algerines as punishment for his having formed a plot to escape from captivity. This man told Brillat-Savarin that, despite lacking a tongue (and, obviously, the taste buds found on the tongue), he had full appreciation of tastes and flavours, but that acid substances produced intense pain for him. For him, the nose clearly supplied more information about flavour than the tongue. This encounter confirmed the view that Brillat-Savarin had developed about the role of smell in flavour. His observations about the contribution that smell makes to flavour predate the scientific pinpointing of retronasal smelling by 160 years.

Taste and smell are so closely fused that we find neurones in the brain which respond, uncommonly, to both taste and smell stimuli, even though the nerves for each sense enter the brain through completely different channels, and are received in completely different parts of the brain. Cells in these brain areas have been shown to be sensitive not only to taste and smell, but also to other sensations such as temperature and texture. We see here the small neural signs of how the brain builds up from separate senses the single sense of flavour. But there are bigger neural signs. When the brain receives signals about taste or about retronasal smelling separately, it is activated as expected: our tongues taste some salt and the taste area of our brain is activated, or our noses get a whiff of thyme and our olfactory system is activated. But when it receives the same signals simultaneously much more of the brain is activated. The construction of the fused sensation of flavour takes a lot of brain power, and it is one of the more complex things that the human brain has to do.

The five senses that contribute to flavour are taste, smell, touch, sight and sound. Each has its separate nerve receptors, and each transmits information to the brain through separate channels. Taste nerves have been shown to be sensitive to five basic tastes. The traditional tastes in Western culture have been salts, acids, sugars and bitter compounds. Asian people have long identified a fifth type of stimulus, the amino acid glutamate, which they link to a perception of meat or savouriness, which in Japanese is called '*umami*'. The taste nerves connect from the tongue to the brain stem, the lower part of the brain that is continuous with the spinal cord and that is responsible for automatic functions such as heartbeat and breathing. This is in sharp contrast with the olfactory nerves which enter the front of the brain as part of the brain's limbic system, where the brain's emotional activity is centred. Taste and smell, though fusing together in the perception of flavour, are anatomically poles apart.

Beyond taste and smell, our sense of flavour is heavily shaped by the feel of food. This sensation is something we prejudge when we take the food to consume it: we compare the expectation of how it should feel against the actual sensation in the mouth as part of judging the mouthful. If the crisps are not crisp, or the gravy is not liquid, or the lettuce is soggy, we are immediately disappointed. There are several kinds of somatosensory receptors, the nerves that tell us the feel of a mouthful of food, in the skin of the lips and cheeks and inside the mouth on its membranes and on the tongue. The receptors inside the mouth are similar to those in other parts of the body, except that they are present in very high density. Obviously, the mouth cares a great deal about the feel, the texture, the temperature and the potentially noxious quality of what it takes in. People who lose their sense of smell and so of flavour tend to seek foods with exciting textures to keep their interest in food active.

Temperature-sensitive receptors in our mouths and elsewhere in our bodies respond to the warming and cooling of the skin. Pain-sensitive receptors respond to mechanical, chemical or temperature stimuli. They can tell us of a wide range of painful sensations: the pricking of a fish bone, burning heat, the heat of chilli pepper, sticking pain, aching pain, in fact all the varieties of pain you can think of. Touch-sensitive receptors can sense small roughnesses, depressions, bulges, stretching of the interior membranes of the mouth. Our pressure-sensitive receptor nerves are extremely responsive to pressure such as vibrations of rapid tongue and jaw movements. All these sensations are communicated through the trigeminal nerve (the nerve in the face that transmits pain) to the brain stem and so to the thalamus: the brain's hub for relaying signals from all the senses except the sense of smell.

The look of food and drink matters too. It is not simply that the sight of food makes us salivate; the sight of food creates expectations of flavour which change our perception of how they actually taste and smell.

An experiment was conducted in Bordeaux in 2001 with red and white wines to test the influence of colour on the perception of flavour. The tests were performed in two stages. In the first stage, the subjects, all students at the Faculty of Oenology at the University of Bordeaux were invited to taste and describe red and white wines. They correctly applied to red wines terms typically applied to red wines, and correctly applied to white wine the usual terms for white wines. A week later they were invited back to taste again, unaware that they were now tasting exclusively white wines, half of which had been coloured red with a flavourless colouring. In this second round, they applied the descriptors for red wines to the white wines that were falsely coloured red, though the flavour was of white wine. They had been fooled by the colour into expecting and perceiving a red-wine flavour. This experiment is much quoted by those suspicious of wine buffs; it seems to confirm the suspicion that wine buffs pretend to discern more subtleties in wine than the rest of us. It would however be fair to say that (as in so many experiments of this kind) the subjects were first-year students, and the results say more perhaps about the sheepish anxiety of first-year students than about the fallibility of wine buffs.

Finally, the sound of food and drink in the mouth – its crunch, its snap, crackle and pop, the satisfying *glou-glou* of wine – these have a part to play in flavour. These have in turn their own separate auditory channel into the brain.

Five senses fused into one sensation; five nerve channels being brought together into a single image of flavour in the brain – it is a prodigious act of synthesis. In creating the illusion of flavour, our brains do more than add together the sum of these five senses: they modify the sensations from one sense in the light of sensations from other senses. For example, sweet tastes combining with sweet smells produce a more intense effect than the sweet taste on its own, or the sweet smell on its own. This is called a superadditive effect. That explains why, when smell is taken out of the flavour mix, the effect is a much less intense and rounded experience. Superadditivity, the mutual enhancement of the senses, applies to other experiences than flavour. It explains why there is an extra pleasure to bringing the sense of smell into play alongside sight and sound, as all the senses are then enhanced.

The interplay of the senses produces some interesting twists. A sweet taste makes food feel thicker and more viscous, whereas sour tastes make food feel less viscous. Thicker, more viscous food appears to give less flavour through retronasal smelling than more liquid food. Coloured food and drink are more likely to seem smelly than colourless things. The colour of food and drink creates expectations about what they will taste like, expectations which can enhance the flavour if they are met, or reduce the flavour if they are confounded. The heat of peppers reduces sensitivity to smell and taste. A strong smell conversely reduces sensitivity to irritation. Heat, in turn, increases the stinging quality of spicy food, and coolness decreases it. A food whose temperature changes in the mouth, ice cream for instance, has complex and delightful effects on flavour as it melts. It is all quite a conjuring trick, and clever chefs have been inventing more ways of playing on our sense of flavour that make use of this.

Brillat-Savarin, a man who relished his food to the full, concludes with the assertion that flavour is the one of our senses that:

> gives us the greatest joy:
> (1) Because the pleasure of eating is the only one which, indulged in moderately, is not followed by regret;
> (2) Because it is common to all periods of history, all ages of man, and all social conditions;
> (3) Because it recurs of necessity at least once every day, and can be repeated without inconvenience two or three times in that space of hours;
> (4) Because it can mingle with all the other pleasures, and even console us for their absence;

(5) Because its sensations are at once more lasting than others and more subject to our will;
(6) Because, finally, in eating we experience a certain special and indefinable well-being, which arises from our instinctive realisation that by the very act we perform we are repairing our bodily losses and prolonging our lives.

CHAPTER 10

The Great Stink

'In Köln, a town of monks and bones,
And pavements fang'd with murderous stones
And rags, and hags, and hideous wenches,
I counted two and seventy stenches,
All well defined, and several stinks!
Ye Nymphs that reign o'er sewers and sinks
The river Rhine, it is well known,
Doth wash your city of Cologne:
But tell me, Nymphs! What power divine
Shall henceforth wash the river Rhine?'

Samuel Taylor Coleridge – *Köln* (1828)

In June 1858, a series of questions was put by Members of Parliament to the Chief Commissioner of Works about the state of the River Thames in London. 'The Great Stink' of 1858, as newspapers at the time called it, was underway: the Thames had become so contaminated by the effluent of London that the Houses of Parliament, on the edge of the river, had become barely habitable during the summer months. The curtains on the river side of Parliament had to be soaked in chloride of lime to counteract the stench. Serious consideration was given to moving Parliament to Oxford or St Albans.

State of the Thames – Question. Hansard 11 June 1858

> Mr BRADY said, he wished to put a question to the noble Lord the Chief Commissioner of Works regarding the state of the River Thames. It was a notorious fact that honourable Gentlemen sitting in the Committee rooms and in the Library were utterly unable to remain there in consequence of the stench which arose from the river; and he wished to know if the noble Lord has taken any measures for mitigating the effluvium and discontinuing the nuisance.
>
> Lord JOHN MANNERS said, he was very sorry to tell the honourable Gentleman that the River Thames was not in his jurisdiction and therefore not under his control.

Ironically, the stench of the Thames was in part a result of improving sanitation. Until the early 19th century, the effluent of London went largely into cesspools. The Thames, though mildly polluted since Roman times, was still clean enough at the beginning of the century for salmon to be caught plentifully in central London. Lord Byron swam in the stretch of river just downstream from Parliament. Fifty years later, swimming was out of the question. The flushing lavatory, which first appeared in London around 1810, had become increasingly popular and was rendered practically compulsory in 1848 when the use of cesspools was forbidden. The Thames then received directly that which had previously been deposited in countless cesspools. To deal with the sewage problem, the Metropolitan Board of Works was set up in 1835, but its powers were inadequate and nothing effective was done.

Then on 30 June, members of a House of Commons committee suddenly rushed out of their room in the greatest haste and confusion, 'foremost among them,' reported *The Times*, 'being the Chancellor of the Exchequer [Disraeli], who, with a mass of papers in one hand and with his pocket handkerchief clutched in the other and applied closely to his nose, hastened in dismay from the pestilential odour ...; Mr Gladstone also paid particular attention to his nose.' That settled matters. Great men, whose *laissez-faire* principles had prevented them from interfering in the natural working of London life – including a succession of epidemics of cholera – could not stand unflinching against so personal an attack as the Great Stink. They rushed a Bill through Parliament to give the Metropolitan Board of Works the powers that it needed to abate the nuisance. Within seven years London was provided with 82 miles of sewers, which intercepted all the sewage flowing into the Thames and redirected it to pumping stations well to the east of London, where it was discharged on the ebb tide.

The Great Stink represented something of a watershed between the ancient world, which accepted stenches as a fact of life, and the modern world of sanitation, deodorising and public hygiene. You might have supposed that smells have been a constant throughout history, that roses and dung smelled the same to an ancient Roman as to us today. No doubt so, but the significance of smells has changed over time. As markers of health, social standing and religious orthodoxy smells have changed character.

The city had since ancient times accepted as natural some smells that we would today be offended by. Horses and donkeys were the means of transport, and their dung was a fact of street life. Milk was provided by cows and goats kept in ordinary homes. Without chimneys to draw smoke up and away, smoke from domestic fires lingered in the houses and streets. Some urban smells, however, were less readily accepted. Two professions were particular stinkers from the earliest times. Ancient laundries used a range of pungent substances to degrease and clean the fabrics – notably stale human urine, pigs' dung, and sulphur (there were public pots in the streets of ancient Rome to collect human urine for laundries, and the traffic in urine was important enough to be taxed). As late as 1935 in England, laundries of this kind were still at work, using 'old lant' (stale urine) as a cleaning agent and trampling pigs' dung into the cloth. Indeed when soap was finally introduced into laundries and took over from urine, dung and sulphur, some found its unfamiliar smell to be offensive.

Tanneries were responsible for another range of urban stinks, so powerful that they were invariably located on the outskirts of town. Most of the individual processes for turning animal skins into cured leather had their own stench to offend the neighbours, and to keep them informed of the sequence of events in the tannery. To remove hair from the skin, the skin was soaked in urine or an alkaline lime mixture or, more simply, it was allowed to putrefy for a period before being dipped in a salt solution. Once the hair was removed, tanners would 'bate', or soften, the skin by pounding dung into it or soaking it in a solution of animal brains. Bating was a fermentation process which relied on the enzymes produced by bacteria found in the dung, and like other fermentation processes – those for cheese, salami or beer, for example – it guaranteed a good ripe smell in the finished product. Tanning the leather usually involved oak bark, another powerful effluvium. Finally, left-over leather was boiled up to make glue. The smells from the tanneries thus came drifting over the ancient city in a sequence of putrefying flesh, animal dung, urine, oak bark and stinking glue. The neighbours would have been gasping for breath.

The Roman wit and writer of epigrams Martial[24] has some excellent commentaries on smells. Here is his ripe insult of a woman named Thais, invoking many of the fouler stinks of ancient Rome:

> Thais smells worse than a grasping laundry-man's long-used urine pot and that, too, just smashed in the middle of the street; than a he-goat fresh from his amours; than the breath of a lion; than a hide dragged from a dog beyond the Tiber; than a chicken when it rots in an abortive egg; than a two-eared jar poisoned by putrid fish sauce. In order craftily to substitute another smell for such a reek, whenever she strips and enters the bath, she is green with depilatory, or is hidden behind a plaster of chalk and vinegar, or is covered with three or four layers of sticky bean-flour. When she imagines that by a thousand dodges she is quite safe, Thais, do as she will, stinks of Thais.

Urban life was, however, not much smellier than country life, until cities grew too big with the industrial revolution to handle sewage and rubbish, and until coal smoke created smog. Before then, it was just as likely that city dwellers would consider country smells to be foul, as that country visitors to the city would be appalled by urban stinks. Aristophanes, the ancient Greek writer of comic plays, has a character define the difference between himself and his wife in terms of the smells of country and town: 'I belonged to the country, she was from the town ...

On the wedding day, when I lay beside her, I was reeking of the dregs of the wine cup, of cheese and of wool. She was redolent of essences, saffron, voluptuous kisses, the love of spending.'

Perfumes were a feature of life from the earliest civilised times. They had different meanings from the perfumes of modern times. Perfumes then were more of a public affair and not, as now, a matter of personal decoration; they were for both sexes equally and they had little of the erotic significance of today. From the times of the Egyptian Pharaohs, perfumers were part of religious rites. The Egyptians believed that the gods sweated perfume, and that embalming the dead, mummifying them, and burning incense for them gave them the immortality of the gods. It was believed that the dead could actively use incense smoke as a ladder to the gods. An inscription on a Pharaonic tomb read: 'A staircase to the sky is set up for me that I may ascend on it to the sky, and I ascend on the smoke of the incense.'

In ancient Egypt, perfumers were priests and priests were perfumers. The formula for a scent called '*kyphi*' is engraved in hieroglyphics on the walls of a temple. Egypt was still exporting fragrance throughout the ancient world in the first century BC. Its celebrated perfumes included the '*metopion*', an oil of bitter almonds and unripe olives that served as the medium for fragrances derived from cardamom, myrrh, turpentine, galbanum, honey, wine, rush and balsam seed. With the annexation of Egypt by the Roman Emperor Octavian, Rome took over its trade routes and gained access to a plethora of aromatic products.

The Greek gods too were scented, and their love affairs gave birth to fragrant flowers. In Greek mythology, the narcissus flower was created by Hades, god of the underworld, to seduce Persephone; Homer described its scent as 'so sweet that all heaven and earth laughed with pleasure'. The frankincense shrub was created from the body of a king's daughter seduced by the sun. Mint was once the cast-off lover of Hades. The laurel bush was born of a nymph beloved by Apollo.

The ancient Greeks and Romans were familiar with the Egyptian practice of embalming, but considered it an alien tradition. In Greece and Rome, corpses were washed and anointed with perfume. The couch on which a body was laid was sometimes strewn with flowers and, in Rome, a resin-scented cypress branch marked the house of the deceased. Incense would be burnt in the house and along the route of the funeral procession to propitiate the gods. Nero was said to have burnt more incense than Arabia could produce in a year at his wife Poppaea's funeral.

After being carried in procession through the streets, the dead were either burnt or buried. In the former case, the funeral pyre would be made of fragrant woods to which scents were added as an offering to the gods and perhaps also, more realistically, to mask the stench of death. Martial, who proves to be a very quotable writer on ancient smells, has an epigram describing the aromatic preparations for a funeral, an epigram with a sting in its tail: 'While the well-heaped pyre was being laid with papyrus for the flame, while his weeping wife was buying myrrh and cassia, when now the grave, when now the bier, when now the anointer were ready, Numa wrote me down as his heir – and got well.'

The mythological model for cremation was the phoenix. Herodotus reports (though he is not sure himself whether to believe the account) the Egyptian legend that the dying phoenix would be plastered with aromatic myrrh by its offspring, and then cremated. The legendary bird was supposed then to be born afresh from the ashes. Similarly the process of cremating people was one of purification, whereby the dead were released from their material bodies and transformed into pure essence, ready for their new ethereal existence in the afterlife.

These costly funeral rites were only for the wealthy. Martial wrote an epigram about a runaway slave called Zoilus who scrapes a living stealing perfumes from funerals: 'The unguents and cassia, the myrrh that smells of funerals, and the frankincense half-burned that you snatched from the midst of the pyre, and the cinnamon you have grabbed from the bier of death – these, rascally Zoilus, surrender out of your foul pocket.'

Perfumes then were for public pomp as well as private titivation. Darius III, king of Persia in 330 BC, had in his court fourteen perfumers and 46 garland makers. In the 3rd century BC, the Egyptian king Ptolemy Philadelphus held a parade in Alexandria which boasted camels loaded with spices, boys carrying frankincense, myrrh and saffron on golden dishes, a giant figure of Bacchus pouring out libations of wine, golden-winged images of Victory bearing incense burners and innumerable garlands of flowers. At the games held in Greece at Daphni in the 2nd century BC, 200 women sprinkled everyone with perfumes out of golden pitchers. During the games themselves, 'on the first five days everyone who came to the gymnasia was anointed with a saffron perfume … And in a similar fashion in the next five days there was brought in essence of fenugreek, and of amaracus, and of lilies, all differing in their scent.'

The homes and possessions of the well-to-do of ancient Greece and Rome were perfumed with the same care that their persons were. The walls of rooms would be daubed with perfumed unguent and the mosaic floors sprinkled with scented water and strewn with flowers. When the weather was cold, fires of aromatic wood would keep the homes of the wealthy both warm and fragrant. Cushions might be stuffed with dried herbs; clothes perfumed with incense and stored in chests of fragrant wood; bath and pool water might be perfumed as well. Even domestic animals such as dogs and horses might find themselves anointed with their owner's favourite scent.

At the other extreme the homes of a poor family would smell very different. A sense of what this smell would be like is conveyed by Martial's description of a poor family's household goods:

> a cracked chamberpot was making water through its broken side. That there were salted gudgeons too, or worthless sprats, the obscene stench of a jug confessed – such a stench as a whiff of a marine pond would scarcely equal. Nor was there wanting a segment of Tolosan cheese, nor a four-year-old chaplet of blackened penny royal, and ropes shorn of their garlic and onions, nor your mother's pot of foul resin, the depilatory of dames.

In antiquity, scents, whether inhaled through the nose or absorbed directly by the body, were regarded as important healing agents. The ancient custom of applying perfumes to the head and chest was not simply an aesthetic practice but also a means of promoting health. 'The best recipe for health,' wrote the poet Alexis, 'is to apply sweet scents to the brain.' Anointing the head with perfume when drinking was believed to counteract the intoxicating effect of alcohol on the brain. Anointing the chest with perfume, in turn, was thought beneficial to the heart. Garlands worn around the head also had the effect of supplying healthful odours to the body. In the case of wounds, perfumes would be applied direct to the injury. A lotion of wine and myrrh was prescribed for burns. Nor was this all superstition – many of these applications would indeed have promoted healing by acting as germ killers.

Aromatic plants in general provided the ancients with a wide variety of curative scents. The Roman writer Pliny the Elder, who was a contemporary of Christ, mentions a number of such aromatic plant-based remedies in his *Natural History*: rue in vinegar was given to comatose patients to sniff as a kind of smelling salts. Epileptics were treated with the scent of thyme. The smell of pennyroyal was held to protect the head from cold and heat and to lessen thirst. The scent of a sprig of pennyroyal wrapped in wool was believed to help people suffering from recurrent fevers, while the smell of pennyroyal seeds was employed for cases of speech loss. Mint scent was thought to refresh the spirit, and was commonly used to ease stomach aches. The smell of anise was used to ensure an easy childbirth, to relieve sleeplessness and hiccups and, when boiled with celery, sneezing. Fumigation with bay leaves in turn was believed to ward off the contaminating odours of disease. Healing smells were believed to emanate from some foods as well. The fragrance of apples, much appreciated by the ancients, was held to reduce the effects of poison. The smell of boiled cabbage soothed headaches.

With the rise of Christianity in the Roman Empire in the 4th century, the use of perfume began to fall into disfavour. Even incense, which was to return to favour in Christian churches, was at first condemned as part of the trappings of idolatry. The early Christian antipathy to incense is hardly surprising, considering that the standard test of loyalty to the Roman Emperor was to burn incense before his image, and many Christians were put to death for refusing to do so. Personal use of perfumes was also considered a frivolous luxury tending towards sensual lusts. Indeed, in their reaction against pagan sensuality, many Christians even ceased washing themselves and were proud to reek of honest sweat and dirt, the scents allotted to the human body by its Maker. Slowly however, Christianity came round to many traditional practices and beliefs about scents. Thus by the 6th century incense, which had played a major role in the religion of the Romans and Egyptians, had become an accepted part of Christian ritual as a symbol of prayer rising towards God. Fragrant flowers and odours figured in many early Christian legends, serving as symbols of virtue or the miraculous signs of grace, and a new Christian idea about smell was added to these ancient traditions – that of the odour of sanctity.

Early saints like Simeon Stylites, an extreme ascetic in the 3rd century AD who earned his sainthood by sitting for 37 years on top of a column near Aleppo, were believed to give off the sweet fragrance of holiness. When he fell ill with fever on his pillar, he was supposedly surrounded by an incomparably sweet scent which grew in intensity until he died. For saints already dead, the sweet fragrance that was encountered when their bodies were disinterred was one of the miracles that scored towards their status as saints. The sweet odour of sanctity stood in sharp contrast to the sulphurous stink of the devil and his agents. Many a woman proved herself a witch by her foul stink.

While the science of perfume stagnated in the Christian West, it advanced in the Arab world. As the heirs of ancient knowledge in the field, Arabs played a decisive role in the evolution of perfume with the development of the techniques for distilling aromatic material to extract the concentrated fragrance. Europe benefited from these advances, and the first alcohol-based perfume appeared in Europe in the 14th century: *The Queen of Hungary's Water*, which was based on wine and rosemary.

In the Middle Ages and into the Renaissance, houses throughout the Western world were smelly places, and their inhabitants fought smell with smell. Animals frequently shared houses with humans, and in most dwellings horses and cattle were stabled in the same buildings as people. Sewage went into cesspools, rather than drains, and an effective chemistry of disinfection was centuries away. Smoke from cooking fires was a powerful ingredient in the mix, since most houses did not have chimneys to draw off the smoke until the 16th or 17th centuries. Smoke was seen by some as a good antidote to some of the other effluvia of the home. When chimneys replaced the traditional hole in the roof, some felt the effect would be bad for their health. An Elizabethan rector noted that colds had been rare before the introduction of chimneys, 'for as the smoke in those days was supposed to be a sufficient hardening for the timbers of the house, so it was reputed to be a medicine to keep the goodman and his family from the quack.'

Strewing herbs were mixed with rushes or straw as a covering for the floor, both to fragrance the room as they were trodden on, and to discourage mice, rats and fleas. Thomas Tusser, a poet and farmer in Tudor times, wrote *Five Hundred*

Points of Good Husbandry, which lists the typical strewing herbs of an English house in the 16th century: 'Basil, Balme, Camomile, Costmary, Cowslops, Daisies, Fennell, Germander, Hops, Lavender, Santolina, Marjoram, Mawdelin, Pennyroyale, Roses, Mints, Sages, Tansie, Violets, Winter Savery'. It seems a delightful meadow underfoot, but we have the testimony of the Dutch humanist Erasmus at the start of the 16th century to give a touch of reality to the scene. He wrote of visiting English houses in which 'the floors are made of clay and covered with marsh rushes, constantly piled on one another, so that the bottom layer remains sometimes for twenty years, incubating spittle, vomit, the urine of dogs and men, the dregs of beer, the remains of fish and other nameless filth. From this an exhalation rises to the heavens which seems to me most unhealthy.' Erasmus's take on the unhealthiness of bad smells reflects the perception, common until the medical advances of the second half of the 19th century, that smells communicated disease.

As well as their use as strewing herbs, rosemary, lavender and other 'astringent' herbs have been burned as fumitories to keep away disease and to clear rooms that had harboured disease. Mixed incenses were often made into pastilles or tablets for burning. Sir Kenelm Digby, in his *Choice and Experimented Receipts in Physick and Chirurgery* of 1668, wrote: 'To make a perfume to burn: take a half a pound of damask rosebuds, Benjamin three ounces beaten into a powder, half a quarter of an ounce of Musk, and as much of Ambergris, the like of Civet. Beat all these together in a stone mortar, then put in an ounce of Sugar, and make it up in Cakes.' This recipe was published three years after London's great plague, and is yet another sign of the close connection made by people for many centuries between smells and illness: both as the causes and as the remedies for illness.

It was universally believed that the bubonic plague was conveyed by corrupt air, and that strong smells would protect against its infection. Municipal authorities of the time had bonfires of aromatic wood burnt in the streets to purify the atmosphere. People fumigated their homes with incense, juniper, rosemary, vinegar and gunpowder. Even burning old shoes was thought better than nothing, while, for extra olfactory protection, some families kept a goat in the house. Thomas Dekker writes that during the Great Plague of 1665 'the price of flowers, herbs and garlands rose wonderfully, in so much that rosemary which had wont to be sold for twelve pence an armful, went now for six shillings a handful'. In his *Journal of the Plague Year*, Daniel Defoe describes a gravedigger whose preventative against infection consisted of 'holding garlic and rue in his mouth, and smoking tobacco'. The gravedigger's wife, who worked as a nurse, protected herself by 'washing her head in vinegar … and if the smell of any of those she waited on was more than ordinary offensive, she snuffed vinegar up her nose'. Defoe describes a well-attended church service during an outbreak of the plague, writing that 'the whole church was like a smelling bottle; in one corner was all perfumes; in another, aromatics, balsamics, and a variety of drugs and herbs; in another, salts and spirits.' Physicians often wore 'nosebags' filled with herbs and spices over their noses when visiting patients.

In non-plague times, too, people made use of personal smelling devices to make the air they breathed smell good and to ward off unhealthy smells. Pomanders, usually carried in metal cases with piercings to let the scent out, were carried as pendants, cane tops, or weights on belts. They were made of resinous or gummy substances mixed with herbs and spices. Some pomander cases had sections for several different scents, as well as a compartment for a sponge soaked in aromatic vinegars. For those who could not afford the expensive resin pomanders, bouquets of aromatic herbs and flowers called nosegays or 'tussie-mussies' served the same purpose. Ophelia's flower speech in *Hamlet* shows that the choice of scented flowers could conceal meanings: 'There's rosemary, that's for remembrance; … and there is pansies, that's for thoughts … there's rue for you …; there's a daisy; I would give you some violets, but they withered all when my father died.'

In 1660, King James II created the post of Royal Herb Strewer, to which he appointed a Bridget Rumney as 'garnisher and trimmer of the chapel, presence and privy lodgings' of the king and queen. She received a salary of £12 for her annual service to the king, and the same again for her service in attending to Queen Catherine's private rooms and apartments. The post became more ceremonial over the years, and by the time of King George III, his Royal Herb Strewer, Mary Dowle, was appointed to lead his coronation procession. For his coronation in 1820, George IV appointed an old friend, Anne Fellowes, to the post, and she and her six attendants scattered flowers and herbs along the carpet of Westminster Abbey. She applied again for the job on the occasion of the coronation of William IV, but this time her services were not required. Neither Queen Victoria nor later monarchs have appointed a Herb Strewer for their coronations, but the Fellowes family to this day lays claim to the position for the eldest unmarried daughter.

Before the 19th century, bathing was a rarity, and seen as a threat to the health of the body by rendering it moist, soft and liable to invasion by the cold. It was reported of Queen Elizabeth I, for example, with a degree of astonishment at the risks she took, that she took a bath once a month 'whether she need it or no.' King Henri IV of France, finding that one of his ministers had just taken a bath, insisted on postponing his summons to court until the next day: 'He orders you,' according to the note sent on the King's behalf, 'to expect him tomorrow in your nightshirt, your leggings, your slippers and your night-cap, so that you take no harm from your recent bath.' Scented soaps were unknown until the 18th century, when for the first time Castile soap was mixed with aromatic essences. Up to that time soaps were made of whale oil or tallow and potash, and were often coarse and foul smelling.

There were, then, plenty of ambient and personal odours to mask. Perfumes that bear some resemblance to the commercial perfumes of today had begun slowly to emerge, in the first place as 'waters' with primarily a medical purpose. The first alcohol-based perfume appeared in Europe, as we have seen, in the 14th century – *The Queen of Hungary's Water*. It was considered a universal panacea, protecting one from everything, including the plague. '*Water Imperiall*' was used for wounds; aquae vitea for healing; even Eau de Cologne started its life as a plague preventive. Perfumes became progressively more popular with wealthy Europeans. Scents of animal origin were particularly popular – musk, civet and ambergris. They were seen as aphrodisiac, so when a character in *Much Ado About Nothing* rubs himself with civet 'it is as much as to say, the sweet youth's in love'. The pleasure that the wealthy took in perfume reached a height in the 17th century, when among the well-to-do everything from letters to lapdogs was scented. At the court of Louis XIV in Versailles, perfume was used in many accessories – bags, fans, handkerchiefs, clothes, wigs, beads and scented gloves. Louis himself, nicknamed 'the sweet-smelling', enjoyed watching his perfumer Martial (a happy reminder of the Roman writer of epigrams with a nose for smells) prepare perfumes. The Prince of Conde assisted in perfuming the king's tobacco with orange blossom, rose, jasmine, musk or civet. To entertain themselves, courtiers made various perfumed products, following the example of the Duchess of Aumont, wife of the Marshal of France, who put together a perfume called '*à la Maréchale*' based on iris, coriander, cloves, sweet flag and nut grass.

The history of smells was, however, heading for a revolution. By the end of the 18th century, changing ideas of public and personal hygiene were beginning to drive out the traditional acceptance of strong smells and to usher in a sanitised and deodorised world. By the early 19th century, movements for sanitary reform began to grow in the cities of Europe. With the multiplication of factories and the bloating of urban populations, the problem of sewage disposal became overwhelming, culminating in 1858 in the Great Stink in London. Sanitary reform was needed, and was made more pressing by the cholera and typhus epidemics in many cities during the first half of the century, epidemics which were suspected of being spread by the smell of effluent. The great urban clean-up followed hot on the heels of the Great Stink: flush toilets were slowly adopted everywhere, and every town acquired its collective sewage system.

With the revolution in civic cleanliness there came a revolution in personal cleanliness, powered by dramatic advances in medical and chemical science. The shift that can be traced back to the Great Stink has seen the progressive eradication of huge ranges of public smells; the growth of perfumes for personal adornment and the commercialising of fragrance on a vast scale; the chilling and packaging of perishable food to eliminate many doubtful smells in shops; the emergence of entirely synthetic smells; and recently the development of public perfuming ('ambient scenting') reminiscent of the ancient Egyptians and Romans but on a scale that would beggar their understanding. Many natural smells have more or less disappeared from the lives of most people, while the smell of synthetic things unknown to the old world – petrol, disinfectants, exhaust, paints, plastic, Play-Doh – fills our noses.

Modern ideas about hygiene and sanitisation only emerged in the second half of the 19th century. Before, it had been desirable to leave one's skin unwashed to prevent it from being invaded by external effluvia; now it was seen as necessary to wash one's skin so as to allow it to release corrupt internal fluids. This sanitising drive was given scientific backing by Pasteur's discovery that it was germs that communicated diseases such as cholera and typhus. Smells were no longer the carriers of disease, but equally they were no longer protection against disease. Hygiene was now the watchword.

As the upper- and middle-classes began to wash frequently and to sanitise their homes, they grew more sensitive to the bad smells of the lower-class people who did not wash and clean. For the poor, the old standards of personal cleanliness held good until the turn of the century and beyond. As a result, the acute class-consciousness of the time was given a new, olfactory twist. It became a commonplace that 'the lower classes smell'. These are the words used by George Orwell when he wrote in 1937 about the lives of poor people in industrial Lancashire and Yorkshire. His book *The Road to Wigan Pier* is an agonised account of the hopelessness of poverty, in which Orwell desperately wants to champion the poor but finds his fastidious nose getting in the way:

> Here you come to the real secret of class distinctions in the West – the real reason why a European of bourgeois upbringing, even when he calls himself a Communist, cannot without a hard effort think of a working man as his equal. It is summed up in four frightful words which people nowadays are chary of uttering, but which were bandied about quite freely in my childhood. The words were: The lower classes smell ... And here, obviously, you are at an impassable barrier. For no feeling of like or dislike is quite so fundamental as a physical feeling. Race hatred, religious hatred, differences of education, of temperament, of intellect, even differences of moral code, can be got over; but physical repulsion cannot. You can have affection for a murderer or a sodomite, but you cannot have affection for a man whose breath stinks – habitually stinks, I mean.

Smells gave an unpleasant edge to the prevailing class consciousness of the time. They also played a part in the prevailing idea of racial superiority of the time. Hitler, and others like him who believed that the human race should be improved by breeding and culling, framed their ideas in terms of racial hygiene. In *Mein Kampf*, Hitler characterised the Jews as unclean agents of racial pollution:

> The cleanliness of [Jews], moral and otherwise, I must say, is a point in itself. By their very exterior you could

> tell that these were no lovers of water, and, to your distress, you often knew it with your eyes closed. Later I often grew sick to my stomach from the smell of these caftan-wearers … All this could scarcely be called very attractive; but it became positively repulsive when, in addition to their physical uncleanness, you discovered moral stains on this 'chosen people'.

But the Nazi campaign to clear Europe of the Jews created its own great stink, which the Nazis had not anticipated, and which leaked out despite their efforts to keep it secret. 'During bad weather, or when a strong wind was blowing,' admitted Rudolf Höss the commandant of the Auschwitz death camp, 'the stench of burning flesh was carried for many miles, and caused the whole neighbourhood to talk about the burning of Jews, despite official counter-propaganda.'

Personal hygiene and racial hygiene were powerful ideas from the middle of the 19th century for at least a century. Paradoxically, though, the rise in personal cleanliness was accompanied at first by a decline in the use of perfumes. No longer attributed any protective qualities by the medical profession, perfumes were instead deemed to clog the pores, or to enfeeble through their heavy vapours. The great 19th-century sanitary reformer Edwin Chadwick considered that 'all smell is disease'. Perfumes were taken out of the apothecary or chemist, and relegated to the category of cosmetics, an accessory to female beautification. They also shifted away from the more animal scents towards lighter, floral scents. Musk became a shocking perfume to wear in Victorian Britain. For the first time perfumes became effectively feminine. Traditionally, men and women had worn the same perfumes. During the

19th century this changed radically. Women wore sweet, floral, feminine scents. Men did not lest they be thought effeminate. Masculine perfumes had to bear a different label – cologne, aftershave – and had to have the husky connotations of leather, musk, tobacco and alcohol.

With the growth of organic chemistry towards the end of the century, perfume was liberated from its natural origins. Natural ingredients, lovely as they might be, were slow to extract, difficult to produce in quantity, and impossible to control exactly in mass production. In 1868, the chemist Perkin synthesised a molecule that smelled like the main odorant of the tonka bean – coumarin. This artificial product, evoking the smell of mown hay, was present for the first time in a perfume that became famous, made by the celebrity perfumer Houbigant – *Fougère Royale*. Soon afterwards, another chemist Reiner launched the industrial production of vanillin, the cheap synthetic version of costly vanilla, and another chemist created artificial musk. In 1898, the German chemist Tiemann synthesised a molecule, ionone, that smelled like the main odorant of violets. The perfumer François Coty used ionone to launch the perfume *L'Origan* in 1905, which was revolutionary at the time, and is still on the market.

Synthetic ingredients were at first looked down on by luxury perfumers, who considered them to be vulgar debasements of natural fragrances. But soon synthetic products came to be happily accepted as the tools of creative perfume making. New molecules could be created with unexpected and novel smells; perfumes that exaggerated nature could be created; perfumes that stood apart from nature and expressed the hidden mysteries of womanhood could be created. Not to mention the advantages in cost, quality control and mass production.

From the 1970s, a pernicious new factor took hold of the perfume industry. Under the growing influence of marketing, new sales techniques appeared which gave much more importance to advertising and communications than to the perfume itself. The trend was to produce high volume quickly. Many perfumes were copies, or copies of copies. Their brand names reveal the associations they are intended to evoke – erotic, neurotic, narcotic. *Opium*; *Magie Noire*; *Obsession*; *Yvresse*; *Addict*; *Hypnose*. So many perfumes are now developed and launched that the great perfumer Edmond Roudnitska called it 'an olfactory cacophony'.

It is now a mass market and on a scale unimaginable in former times. This reflects, of course, the growing affluence of people at all levels of income and social status. We can all pretend to the status of Louis XIV. It reflects, too, the bizarre link-up of perfume and travel, brought about by the duty free shop. These two former luxuries are no longer for the select few: they are for the millions trudging through airport terminals, assailed by mass marketing and bamboozled into believing that they are experiencing the pampered luxury of the elite. Where there are mass profits, there are gross corporations wanting to gulp down the profits and to exclude smaller fry from the market. In place of artisanal perfumers, there are now a few large international corporations which have raised the cost of launching a perfume to the point where smaller perfume houses can only compete at the fringes, where volumes are low. A perfume is no longer created by a perfumer, but by a multi-disciplinary team whose members are charged with choosing the celebrity sponsor, writing the brief for a fragrance house, creating the synthetic ingredients, making the mix, choosing the marketing pitch and calculating the cost.

Recently a new phenomenon has emerged which harks back to ancient Rome. This phenomenon is the machine for pumping artificial scents into the air at corporate events and extravagant weddings. In the fading years of the Roman Empire, the emperor Heliogabalus once held a banquet at which he had tons of rose petals dropped from the ceiling, and some of his guests reportedly died of suffocation. Technology has now given us a safety-conscious way of achieving the same lavish effect. We have entered a brave new world of smell.

CHAPTER 11

Brave New World

Brave New World is a futuristic fantasy, written in 1931 by Aldous Huxley, which imagines a world in which people can gratify their material desires without restraint. It is set in our world, 600 years ahead. However in a foreword to a later edition, the author writes that he may have misjudged how long it would take for reality to catch up with his fantasy. 'It is,' he writes, 'quite possible that the horror may be upon us within a single century.'

Less than a century later, quite a lot of the prognostications in *Brave New World* are firmly installed. In Huxley's fantasy there are test-tube babies, as there are in reality now. From an early age, children are conditioned to desire the material things of life. Everything is disposable: 'Ending is better than mending' is one of the slogans of the Brave New World. God for most people has been replaced by the economics of mass production. There is high-speed travel to distant continents. There are televisions in every hotel bedroom. There is piped synthetic music everywhere. Even – amazing fantasy for 1931 – electric clocks that require no winding and that are synchronised automatically to the second.

Children in Huxley's utopia are encouraged to dislike wild nature but to love expensive country sports and manufactured goods. So they are conditioned, for instance, to hate the country but simultaneously to love all country sports which entail the use of elaborate apparatus. Huxley's notions of suitably materialist country sports were wide of the mark – he had not imagined mountain bikes or fashionable ski gear – but the principle has been applied today with a vengeance. Of all the imaginary utopias and dystopias thought up at that period, Huxley's is the one that has come true.

Some aspects of the fantasy now seem ludicrous. In Huxley's new world, the conditioning of the population is done by an all-controlling, totalitarian State, not by market forces. He failed to conceive of computers and, although his world is rich in information, it is all held in vast card indexes manipulated by lower-caste cloned people. He thought helicopters would become the principal means of transport, and that they would be guided by giant lighthouses dotted around the landscape, a landscape devoid of everything but golf courses, tennis courts and factories. When his hero finally flees the materialist world, he takes refuge in a derelict helicopter lighthouse in Surrey, near Puttenham, a modest village nine storeys high, with silos, a poultry farm and a vitamin D factory.

Scent, in this future world, is one of the material luxuries available on tap to all. All hotel bathrooms offer eight scents for the bath. Public places are scented by a scent organ. Westminster Abbey is remodelled as an entertainment centre, with London's finest scent organ and all the latest synthetic music. Huxley describes the scent organ playing a 'Herbal Capriccio' – arpeggios of thyme and lavender, of rosemary, basil, myrtle, tarragon; a series of modulations through the spice keys into ambergris; and a slow return through sandalwood, camphor, cedar and new mown hay (with occasional subtle touches of discord – a whiff of kidney pudding and pig's dung) back to the simple aromatics with which the piece began.

And lo! It has come to pass as Huxley foresaw. Any of us with cash to spare can hire a scent diffuser, an electric machine which pumps synthetic fragrances into the air. We can choose the scents to reflect the mood we wish to manufacture. To create an 'immersive experience' for our guests, these scent machines can be loaded with multiple scents, and, just like the outpourings of the scent organ in Westminster Abbey, the smells will change as the event progresses. Scents can be programmed to work in conjunction with lighting, audio, video and special effects. The machines have a variable speed blower and variable scent rate control.

Scenting (according to the sales literature) is the new factor in many big events, like weddings. For the ceremony, companies that hire out scent diffusers suggest that you might choose white gardenia, white rose and orange blossom. For the cocktail hour, the fragrances of mojitos, vanilla, grapefruit and lime martinis. For the reception, chocolate and wedding cake aromas. But a glance at the catalogue shows that the choice of scents goes very much further, and that the choices for your wedding could be much more daring. There are almonds, amaretto, apples green and red, smokey bacon, bananas, basil, beef, blackcurrant, boiled cabbage, brandy, bread, bubble gum, cake shop, candy floss, caramel, carrot, celery, cherry, chicken, chocolate, Christmas pudding, cinnamon, coconut, coffee, cola, coriander, curry, fish market, fruit cake, garden mint, garlic, gingerbread, Granny's kitchen, grapefruit, herbs (mixed), herring, honey, hot apple pie, jambalaya, jelly babies, kiwi fruit, lime, liquorice, mango, melon, mulled wine, nutmeg, onion, orange cupcake, passion fruit, peach, pear drops, peppermint, pina colada, pineapple, popcorn, raspberry, rhubarb, rosemary, rum, smoked fish, spices mixed, stir fry, strawberry, sweet sherry, tangerine, tea leaf, Thai curry, toffee apple, tomato, watermelon, whisky, wine oak cask, vanilla, whisky. And that is just the simple fragrances.

There are more sophisticated choices still: alpine laundry powder, aristocrat, baby powder, boiler room, brewery, burning peat, cannon, carbolic soap, Caribbean holiday, Christmas tree, church, hospital clinic, coal face, Cuban cigar smoke, cut grass, Davos, dentist, dinosaur, football dressing room, farmyard, fish market, flatulence, gun smoke, iron smelting, Japanese prisoner of war camp, man of war, mummy (Egyptian), old inn, old river, old smithy, out at sea, phosgene gas, pit ponies, rubbish, stables, star's dressing room, steam train, swamp, sweaty feet, tropical rain forest, urine, volcano, vomit and Yuletide. Self-indulgence up to the limits imposed by hygiene and economics, as Huxley predicted.

It would be an exaggeration to suggest that all these scents are being pumped out at this moment. Some of the fouler concoctions are included in the catalogue to intrigue the customer; not many people are buying flatulence or vomit. But there is a growing market for scents to be pumped out for film previews, for pop music extravaganzas, for catwalk shows, for marketing launches and for theme parks. Though most of the scents are safe, the fouler options are being used for events where they want 'to make the children squirm'.

The modern scent diffuser is a step forward from Smell-o-Vision, Odorama and Aromarama – scent systems used in films to pipe smells to cinema seats. The 1960 film *Scent of Mystery* used piped smells, synchronised with the film soundtrack, to reveal clues to the audience. It worked badly, the scents taking too long to reach the audience and to clear away. The cult film director John Waters later paid homage to Odorama in his film *Polyester*, equipping the audience with scratch-and-sniff cards which conveyed the smells of roses, flatulence, glue, petrol, skunk, new cars and dirty shoes.

Yes, I think I will 'theme' my next family Christmas. I will hire a scent diffuser. We will start, naturally, with a whiff of Yuletide and Christmas Tree. But we can sure that there will be discord in the family before the day is out, and we must make sure the day is appropriately 'themed' with that oft-forgotten scent factor. So the scent diffuser will move on to Mulled Wine and Gunsmoke, Granny's Kitchen and Boiled Cabbage, Christmas Pudding and Flatulence.

'Event theming' is the technical term for what the modern scent organ does. It is however just a small corner of a bigger market. A more substantial chunk of the market is engaged in 'ambient scenting'. This activity is concerned with creating, through fragrance, an environment propitious for buying and selling. Aldous Huxley in *Brave New World* predicted that the natural sciences and psychology would be harnessed to the business of creating a complete consumer society. So we find that the business of selling goods and services is supported by chemistry and psychology, some of it of questionable worth.

The driving idea of this world of ambient scenting is: These Scents Will Make Consumers Spend More. There is plentiful research evidence in support of the idea.[25] Much of it leaves one wondering whether any care was taken to control for factors other than scent, and whether the claims from one limited piece of research are not being applied unthinkingly to a much larger field. Here are some of the statements touted by ambient scenters. In a controlled experiment by Alan R Hirsch MD of the Smell & Taste Treatment and Research Foundation Ltd of Chicago, the amount of money gambled in slot machines in scented areas of a casino were found to be 45 per cent higher than the amounts gambled in slot machines in unscented areas of the same casino. In a controlled experiment, floral scents sprayed around the bedrooms of a hotel resulted in more positive dreams than in unscented bedrooms. A Dutch lingerie retailer, Hunkemöller, working with a behavioural agency called BrainJuicer, found that the smell of chocolate boosted the sales in their retail outlets by 20 per cent. When the odour of baked bread was released in a US supermarket, sales in the bakery section increased threefold. It seems that one has only to sign up for a scent diffuser, making of course a careful choice of the scent that will be right for that particular sales pitch, to beat one's rivals. If the advocates of ambient scenting are to be believed, the world economy needs only to install scent diffusers everywhere to experience massive growth.

Every walk of economic life can find a role for ambient scenting. Funeral directors use scent diffusers to avoid any whiff of death. Chains of coffee shops use them to make the pavements fragrant with roast coffee and cinnamon, and to entice consumers in. Hospitals use them to soothe nervous patients – lavender and eucalyptus are thought to be particularly calming. Car hire firms use them to make it seem that your hired car is fresh from the factory, as if it has only just been unwrapped from its plastic packaging. Airports and coach stations scent their waiting areas and business lounges to reduce the stress levels of passengers. Hotels use hints of citrus and herbs to reinvigorate newly arrived guests and give them the energy to visit the hotel gift shop. Offices use scent diffusers to increase productivity – peppermint is thought to increase alertness, and to stimulate high work rates from office workers. Peppermint is used in gyms too, where it has been established (in a controlled experiment, perhaps) to increase running speed, handgrip strength and the number of push-ups. In Aldous Huxley's words, the horror is upon us.

All this is based on flimsy reasoning. One line of thought is that nice smells create a contented environment, in which people are subconsciously made more receptive to the experience they are paying for – be it shopping or sending a loved one to his last resting place. In the language of marketing consultants, 'building a strong emotional association between customers and brand through multi-sensory marketing translates into higher prices that customers are willing to pay.' Please them through their noses, and charge them through the nose.

Another line of thought amounts to the same thing for diametrically opposing reasons. Research has shown, apparently, that people spend more when they are in an environment of 'warm' scents like vanilla and cinnamon, as opposed to 'cool' scents like peppermint. A warm ambient scent, according to this line of thought, leads people to perceive the environment around them as more crowded with other people. The more crowded the space, the less powerful people feel. In order to re-establish their sense of power and status, the consumer then spends more on high-price and high-status products. One hopes that there are not too many poor saps being manipulated by ambient scenting in this way.

A more down–to-earth line of thinking is also at work to justify ambient scenting. This is that many retailers are fearful that smells will put consumers off, and have worked hard to remove potentially offensive odours ('malodours' is a term much employed in the marketing of ambient scenting). Packaged, canned and refrigerated food, for instance, cannot be smelled, so scenting with odours that resemble agreeable food is introduced to complete the experience. That accounts in part for supermarkets offering their customers coffee from coffee machines placed where the flow of air will waft the aroma most enticingly.

Aldous Huxley would regard the development of ambient scenting as a natural flowering of the Brave New World he imagined in 1931. He would surely have considered it as part of the horror of a consumer world in which the expensive synthetic is manufactured to replace the natural, in order, as he put it, 'to keep the wheels turning.' Perhaps we should take a more relaxed view of this development. Perhaps it is not the ordinary consumer who is being manipulated by ambient scenting, but the unfortunate retailer or hotel owner who is being manipulated into buying expensive gear for fear of losing market share.

Yet, innocuous as it seems, there may be a bigger trend here which is leading to an unforeseen and unwanted new world. We eliminate natural odours; we replace them with synthetic odours that are more reliably delightful; eventually we find the natural smell less satisfactory than the synthetic substitute. The synthetic experience trumps the natural experience. Similarly, we enhance the flavour of strawberry ice cream, and make it just a little sweeter and more berry-like than the original; we end by finding the natural strawberry insipid. We edit out the roughnesses and mistakes in a performance of music and end up preferring recordings to live music. We enhance the colours of flowers in photos; and eventually we find that the natural colours of flowers in the garden are not quite vivid enough. In the same way, we enhance sports with pumped sound, firework displays, action replays and slow motion and we end by finding that an unenhanced sport event is a touch disappointing. And so on – the enhanced experience, touched up with computer-generated colouring, pumped up with extra flavour and fragrance, souped up with deeper sound effects, may eventually mess up our pleasure in natural experiences. It may be difficult to reverse the trend if it proves to be unwanted.

Perfumes have made use of synthetic ingredients for over a century. We no longer worry or even notice. Indeed synthetic ingredients have made possible some scents that would have been impossible to create with purely natural ingredients. Amongst the surprising perfumes now available you may find Cannabis Flower, Clean Skin, Gin & Tonic, Holy Water, Snow, Paperback, Salt Air and Thunderstorm.

Perfume houses do not however come entirely clean with their consumers about the synthetic nature of their luxury products. There is every advantage to a perfume house in refraining from disclosing openly the precise ingredients in their fragrances. For one thing, the competition would create a copy in no time. For another, the consumer might not be able to cope with an ingredient with a name like allyl amyl glycolate (a molecule that gives a blaring pineapple smell). Usually, the PR department of the perfume house lists 'fragrance notes'. Some lists are long; some are short. That is no guide to whether one perfume has more ingredients than another. The fragrance notes are all listed as natural ingredients but that is no guide to whether the ingredients are in fact natural or synthetic. The fragrance notes might include jasmine, which could be an essence of real jasmine, or a mix of synthetic and real jasmine, or several synthetic jasmines with no real essence, or a chemical that is vaguely floral but which the PR people have decided to label as jasmine. There are molecules that smell like more than one 'note', and these may be listed as separate entities: the notes 'precious woods, amber and musk' could all come from one synthetic chemical. Another company might use the same chemical and list it as 'musk' or 'sensual musk' or 'exotic woods'.

There is another way in which the world of traditional perfumes has entered a Brave New World, though Aldous Huxley did not anticipate what it would look like. At the heart of the perfume industry there remains the traditional craft of assembling fragrances in a mix which will give pleasure for hours as ingredient after ingredient evaporates. Perfume-making still depends on one creative person, with expertise and a discriminating nose, who makes a beautiful thing. Around that one creative person, however, there is now a platoon of people whose role is to manipulate the consumer. Huxley, who believed that the consumerist world of the future would be commanded by a totalitarian state, would be disconcerted to find that the real consumerist world of today is commanded by the consumer, and that droves of people would devote their time to persuading the consumer to buy more and pay more.

Perfumes are now very big business indeed. The industry is now estimated to be worth $40 billion a year and that figure is increasing yearly. More than 450 new fragrances are launched on international markets every year. At any time, in most countries, the consumer has over 1,000 brands of perfume to choose from. There are more consumers, and they are being persuaded to own more bottles of perfume, to think not of their single perfume but of their perfume wardrobe, from which they can choose this afternoon's perfume according to the mood. They are being persuaded too that perfume is only one part of the 'scent package', to be used in conjunction with other expensive goods such as lotions, hair mists and body splashes all sharing the same brand.

It is a world in which image has become more important than the smell of the product. The classic revelation of how the marketing of perfume works is the case of Parfums BIC. BIC was a successful brand of cheap, disposable, brightly-coloured things such as biros, razors and cigarette lighters. These products had a smart, up-to-date style, and a no-nonsense approach to exposing how they worked internally. They had successfully invaded the faux luxury market of pens, razors and lighters, and had shown the consumer that he or she did not have to pay a high price for a serviceable product. Now they hoped to invade the luxury world of perfumes in the same way, and show consumers that they could acquire a serviceable product at a fraction of the price. They launched a range of Parfums BIC in 1988. The perfume bottle looked like a disposable cigarette lighter: clear plastic with a brightly coloured top. Each bottle, containing about three quarters of a fluid ounce, was priced at $5, radically cheaper than conventional perfumes which might cost, at the top end of the market, $120 per fluid ounce or more. BIC marketed their product with lines such as '*Des vrais parfums à un prix BIC*' – real perfumes at a BIC price; '*L'argent ne fait pas le bonheur*' – money doesn't buy happiness; and '*Le parfum nu*' – the naked perfume.

The BIC perfume was a failure. Consumers were not attracted. It was removed from shelves two years later at a loss of $11 million. It is now seen as one of the classic failures of marketing. Far from undercutting their rivals, far from cutting out the nonsense in order to give consumers the essential, BIC found that their cheap disposable perfume had simply failed to give consumers what they want. It had cut out nonsense that was in fact essential to the quality that consumers seek from perfume. What consumers want from a perfume is not a nice smell at a reasonable price, but a symbol of their own high value. The perfume industry took the lesson to heart, and has worked hard to make consumers feel that they will be worth more for buying their particular brand. The result is a fantasy world of sophistication and allure, created more by PR than by the perfumer. Increasingly perfumes are sold by association with fabulous people.

Many perfumes are now marketed as if they were created by a celebrity, or if not created by them, certainly created for them and worn in intimate ways by them. Where there is no celebrity, there may well be a fictitious woman or man. Here, chosen at random, are a couple of fabulous perfume characters.

> She – is independent and sassy. She drives her own eye-catching vintage car. She parks outside a nightclub, she sprays a fragrance, she throws her hat to the doorman who knows her well. She has no need to play second string to a man. She's beautiful, almost ambiguously so, she's bold and vibrant, she's a breath of fresh air. She has this air of elegance and complete confidence, yet she's so approachable. Her smile says it all – Yes, I'm independent, fabulous, happy and I have it all.

> He – is sexy and mysterious. He strolls nonchalantly through the city with glam rock elegance. Flashiness goes hand in hand with an ambiguous dark side that often pushes him to the limits of extravagant provocation. He is demanding and aesthetic, and considers art as a core value. He lives in the heart of new technologies, surrounds

> himself with works of art and objects of graphic design, carved from high-tech materials, as precious and as dark as carbon. Urbane and sophisticated, he is an epicurean who burns with a contagious passion for life.

A writer of fiction in the 1930s like Aldous Huxley could surely not, in his wildest fantasy, have conceived of a world in which consumer marketing threw up so monstrous a pair of fictions as these two perfume dummies.

There is a different olfactory luxury, occupying a quite different corner of the world to perfumes, and attracting a quite different model of person. Cigars, like perfumes, are a luxurious smell and a status symbol. But the image projected by cigar smokers is radically different to that projected by people using scent. Cigar smokers are almost exclusively men. Unlike the perfume man, the cigar man is far from being ambiguously sexy, far from being glamorous, far from moving through the city with a lithe suavity. The cigar man, as he emerges from the consumer literature of the cigar world, tends to have heavy facial hair, a significant belly and a hunting cap. He tends to be lumbering rather than lithe. And he is seriously knowledgeable.

He is familiar with the various shapes of cigar – Corona, Pyramid, Torpedo, Perfecto, Panatela, Culebra. He knows how a cigar should be stored in a humidor, how to cut a cigar, the correct choice of lighter (on no account paper matches, gas lighters or – worst of all – scented candles). He is knowledgeable about the correct procedure for lighting a cigar, and the correct moment when the band around it should be removed (after twelve puffs, apparently). He is knowledgeable about how to savour the smoke, to rotate the cigar at intervals, to let the ash accumulate, to appreciate each phase of the smoke from start to 'final'. He is knowledgeable about the choice of alcohol to accompany a smoke. Most important, he is knowledgeable about the qualities of tobacco leaf in his cigar.

Curiously, cigar aficionados talk of a cigar tasting, not of a cigar smelling. Of the requirements of a well-blended cigar, flavour comes first, followed by strength, aroma, complexity (the changing experience as one smokes down a cigar), balance and finish. The experience of smoking a cigar is however an olfactory experience, and the language of a cigar tasting is the language of smells: grassy, spicy, metallic, rich, leathery, earthy and so on. Here is a tasting note for a Nicaraguan cigar, taken from the website of a tasting club: 'earthy and spicy notes are dominant. In the second third, the smoke becomes more rich and complex – earthy and spicy with a shadow of black honey. In the final, the full earthiness is combined with Cuban cedar notes and a powerful dark pepper finish.' The language could suit a perfume review almost as well as a cigar review.

To the cigar connoisseur, one of the pleasures is the variability from cigar to cigar and from leaf to leaf. These are not mass-produced items like perfumes, conceived in a laboratory and manufactured under perfectly controlled conditions. They are handmade from tobacco leaves grown under variable conditions, so that the quality of the leaf will vary from place to place and harvest to harvest. The blending of different leaves is a matter of personal judgment by the cigar maker. As a result there is a delightful individuality to each cigar, which gives full play to the arcane knowledge of a connoisseur.

Tobacco is undergoing a revolution that Huxley did not predict, but that he surely would have felt was in tune with his fantasy of a future of self-indulgence up to the limits imposed by hygiene and economics. Fear of lung cancer has in many developed countries driven tobacco smoking out of public places. Centuries of smoke encrustation and nicotine staining have come to an end. The traditional fug of a pub bar, the traditional sticky yellow-brown of the pub ceiling, the traditional layers of tobacco smog at the end of an evening of drinking and smoking, are gone. Technology has leapt in with a synthetic substitute to fill the gap that has been left. Soon after, big business has swum in to the swarm of small fry to gobble up the biggest profits.

A Chinese research pharmacist called Hon Lik, whose father had died of lung cancer, thought up a high-frequency, piezo-electric, ultrasound-emitting element to vaporise a pressurised jet of liquid containing nicotine. He patented the design for what would become the modern e-cigarette in 2003, and registered an international patent for the device in 2007. Ten years later, there are millions of users worldwide. The big tobacco companies have acquired chunks of this growing market.

With e-cigarettes has come a whole new culture. Consumers of e-cigarettes call themselves 'vapers'. The Oxford English Dictionary's new word for 2014 was 'vape'. Large gatherings of 'the vaping community' called 'vape meets' take place, where vapers demonstrate their 'mods': modifications of basic designs to create a customised design. A subclass of vapers configure their gadgets to produce large amounts of vapour by using low-resistance heating coils. This practice is called 'cloud-chasing', and allows the smoker to blow extravagant, shapely clouds of vapour into the air, far shapelier and more artistic than the smoke rings that old-fashioned smokers used to create from their traditional cigarettes.

At the heart of the e-cigarette trend is the aroma of a myriad of flavours of e-liquid. E-liquid, the vapour puffed out by an e-cigarette, consists mostly of propylene glycol, glycerin and water, with nicotine and flavourings. There are more than 8,000 flavours, and the number is constantly growing: vanilla; custard and cream; cherry; mojito; frozen tundra; pineapple; coconut; blueberry champagne, topped with lime and mint; doughnuts; and so on. And of course nicotine or, for those who prefer it, cannabis.

Part of the attraction for vapers in choosing a very individual flavour is that it is exclusive to them. It sets them apart, expresses their individuality, it separates them from other people and gives them their own atmosphere. In a way it harks back to Renaissance times, when rich people had individual nosegays, pomanders or tussie-mussies which expressed their individuality and allowed them to breathe different air to that breathed by others. The choice of flavour for the e-cigarette also harks back to Renaissance times, and the meanings that were imputed then to flowers or herbs. In a similar way the flavour of e-cigarettes conveys some meaning about the user's lifestyle.

Tobacco is fading out for good medical reasons, but we are losing something. Tobacco has, in Europe and Asia at least, for five centuries been a secular incense. For five centuries people have dwelt on the varieties of leaf – from Virginia, Cuba, the Balkans – the varieties of curing – in sun, over smoke, in heat – the varieties of ways of consuming tobacco – as snuff, cigars, cigarettes, pipe or chewing tobacco. For five centuries we have become sensitive to the aromas of smoking tobacco, from the first sulphurous tang of a match to the last, sour, sodden smell of cigarette butts. This may be a moment to think afresh about the pleasures of tobacco scents. It may be the moment to make tobacco into a communal rather than a solitary pleasure, with

Havana leaf pot-pourris, or Balkan Sobranie marinades for food, or Cameroon leaf joss sticks.

One area of life which has modernised without losing its connection to the natural is the world of wine. Wine may have entered a Brave New World, in which our material desires can be gratified without restraint, in which science is applied to the design of a product which is reliably appealing, but it has not severed its link to tradition. The wine business does not deserve to have fun poked at it for having replaced the real with fantasy. Wine is still made largely by small producers in family estates; it still eschews synthetic additives; it is still largely free of the vice of branding; it is still produced by a slow process of vinification. But winemakers have succeeded in giving their wines more aromatic character, and in eliminating aromatic disasters, by understanding better how the process of wine-making affects the aroma of wine. They have not had recourse to artificial ingredients, as has so much of the rest of the food and drink industry. Wine specialists have also succeeded, more than other specialists, in making the public aware of the subtleties of aroma and flavour, in language that is for the most part down to earth. They may stray for a moment into the high clouds of poetic metaphor, or into the depths of chemistry, but wine drinkers laugh at them and bring them back to earth. If occasionally wine connoisseurs take flight into concepts like finesse and noblesse that seem too French for a pragmatic Anglo-Saxon mind, I can forgive them. I can even acknowledge that we Anglo-Saxons should learn more about finesse and noblesse.

One of the olfactory ideas that we, who are not wine buffs, have to understand is the distinction between bouquet and aroma in wine. Bouquet is the quality of smell of wines that have had time to mature for some years in the bottle, a quality of intensity and complexity that may bear little relationship to the original smell of the grape and to the early stages of fermentation. Aroma is the smell of the grape variety and the smells produced during the course of fermentation. Some would say that it is only white wines which have aroma, and red wines which have bouquet. This is so where whites are drunk young and reds are drunk old. But it does not hold true in the opposite case: a ten-year-old Montrachet, a white wine, is all bouquet, and a Beaujolais Nouveau, a red, is all aroma. This distinction between bouquet and aroma has been made by winemakers and wine drinkers for centuries. One wonders, also, whether it might be applied to other drinks and foods, like cheese, that mature and change character over time.

There are two types of aroma to be distinguished. Primary aroma is the original smell of the grape, its aroma before fermentation takes place. It is the scent of the fruit itself, characteristic of the grape variety or even of an individual cloned vine. Certain 'noble' grape varieties, grapes traditionally associated with the highest quality wines and said to retain their character wherever they are grown, are remarkable both for the character and penetration of their primary aroma. One might suppose that the grape variety would be the main source of the aromatic character of a wine. But it is of course not as simple as that. One variety can smell and taste very different depending on where it is grown. Cabernet Sauvignon, for instance, is grown worldwide, but its aroma varies. Wines made from Cabernet Sauvignon smell of resin in Porto Carras, of cloves in the Guadeloupe valley in southern California, of liquorice in Rioja, of seaweed in wines from the Golfe du Lion coast, of grape stems in the Languedoc and so on. Here

I draw on the great Bordeaux wine expert Émile Peynaud, not being able to claim that I have sniffed and compared so many wines.

The secondary aroma of wine which evolves during the course of fermentation is the result of yeast activity. The phenomenon has been called 'aromatic fermentation'. Subsequently these fermentation odours are refined by benign bacteriological changes which create subtle extra facets to the aroma. In the end the smell of a young wine is a blend of aromas, combining those of the grape variety itself and those produced during the course of fermentation.

The aromatic character gradually dissipates so that after several years of ageing, it practically disappears, giving way to bouquet. In fact, one result of improved methods of making wines is that modern wines always retain some aromatic character even when mature. In the words of Émile Peynaud: 'the art of graceful ageing is that of prolonging the virtues of youth'. Only fine wines have the potential to progress from aroma to bouquet; others are said 'not to bouquet'.

Bouquet is most commonly acquired by wines from temperate regions which are stored in conditions which avoid contact with air as much as possible. The barrels and vats are always kept filled to the brim, and are topped up frequently. Wines of this kind finish up in bottles hermetically stopped with long corks which ensure an airtight seal for twenty years or more. There is another sort of bouquet that comes to wines stored without complete protection from contact with air and the bacteria carried in air. In this case, the wines generally come from a warm region and are either naturally high in alcohol or have been fortified by the addition of brandy. Fortification is a method of preserving wine and protecting it from attack by bacteria, which would ruin it. As high alcohol will kill most bacteria it is no longer necessary to take so much care to keep the wine from contact with oxygen and it can be stored in partially-filled barrels. This method of storing promotes what is known as oxidative ageing. The wine develops a bouquet called 'rancio'. Rancio, one of the very few terms that refers only to a smell, is a buttery/nutty smell of spirits aged in oak that is the signature of many sherries, ports, madeira, marsala, cognac and some sweet dessert wines. It derives from the Spanish for 'rancid' but has a wholly complimentary meaning when applied to wines. Aged in contact with air, these wines become immune to airborne contamination. Most do not improve in the bottle, but neither do they deteriorate after several weeks in an opened bottle.

The care devoted to wine making, with the goal not only of making wines with delicious flavours, but also of making wines whose smell in the glass gives pleasure, is heartening. It suggests that we are capable of giving the sense of smell its proper place in our sensory world, and of putting our sense of smell into words. It suggests that there are areas of life where we can enhance our olfactory pleasure without relying on synthetic chemicals. It leaves me wondering whether such devotion to the natural smell of things could not be applied to other things than wine – to streets or flooring or clothes or transport. We have become more clever at controlling the world of smells, and more creative in making new smells for the pleasure of mankind. Our sensory world has become wider as a result. But we are no closer, it seems, to understanding the meaning of what we smell, nor of capturing in words the extraordinary range of sensation it offers us. Perhaps it is time to unlock this neglected sense.

CHAPTER 12

Noses, Numbers and Wheels

History is spattered with attempts to classify smells and to put them into words systematically. Many of these attempts were based on the hope that smells are orderly, in the way that colours and sounds are orderly. Colour and sound come in wavelengths with a definite hierarchy. The hierarchy of colours is visible in the rainbow; the hierarchy of notes is visible on the piano keyboard. Smells however are less orderly, and attempts at finding a grand order of smells have failed.

More limited attempts at classifying the smells of particular subjects have been made. Some have been very successful; some remain very obscure with no impact on the ordinary person who tries to understand the meaning of everyday smells in ordinary life. I imagine the olfactory world to be like a vast equatorial jungle into which a few very purposeful expeditions have cut their way successfully. The perfume industry has launched the most successful expedition, which has travelled far up some rivers and mapped some territory exactly. The rivers of flowers, spices and animal smell glands have been explored and are now colonised by the perfumers and their chemists and the aromatherapists. Smaller expeditions have pushed up some other rivers – the wine trade, coffee blenders, the beer industry. You will even find a tiny niche occupied by mushroom experts who have explored and mapped the smells of their fungi.

But there are huge tracts that have not been mapped. You will not find a good map of woods, grasses, building materials, fish, meat, textiles, the smells of the street, the smells of the farmyard, marine smells, the smells of the hardware shop or the oil refinery. All these remain unexplored territory. And there are subjects, like tobacco, whose smells are interesting aesthetically and commercially, which have been explored in the wrong way, by people who are searching for taste rather than smell. It is something like the state of medieval map making, in which parts of the world are accurately mapped, and parts show mythological monsters and imaginary mountains.

The fringes of this olfactory jungle are littered with the debris of major expeditions by serious people in search of the complete map of all smells, the ultimate classification which identifies the primary smells and binds them into a general system. These major expeditions seemed to be breaking new ground, and for a time they attracted backers, but then they began to falter and were finally abandoned, leaving a scattering of rusty ideas behind them.

Carl Linnaeus, the great Swedish botanist and zoologist, who in the 18th century formalised the modern system for classifying organisms by genus and species, was one of the first to try putting smells into order. He had already put the entire animal and plant world into order, into hierarchies which allowed everything to be described in terms of its relationship to similar things within the same hierarchy. Now he could apply the same approach to smells. He identified seven main categories of odour, and hoped that every smell would slot into place. In the Latin that scientific thinking of that era relied on, his categories were: *odores aromatici* (laurel smells); *odores fragrantes* (lime or jasmine blossom smells); *odores ambrosiaci* (amber or musk smells); *odores alliacei* (garlic smells); *odores hircini* (male goat smells); *odores tetri* (the smell of solanaceous plants like tomato and potato); *odores nausei* (foul-smelling plants). These classifications were then used to describe the medicinal properties of substances used for treating patients. As a general classification of smells it does not succeed.

There is no place in it for such common smells as smoke or wine or fish or bitumen or bread. Nor does Linnaeus's classification offer much of a basis for the fine discrimination between smells that comes so naturally to us, the subtle discrimination that allows us to distinguish, for instance, grapefruit from satsuma from orange. But such was the authority of Linnaeus that his approach to classifying smells persisted for a couple of centuries.

Some bolder attempts at a general classification of smells were made in the 19th and 20th centuries. A Dutch physiologist called Hendrik Zwaardemaker designed a classification based the seven primary odours that Linnaeus had codified, plus two: ethereal (which covered such smells as ether and beeswax) and 'empyreumatic' (meaning burnt, and applying to such things as coffee and tobacco smoke). One occasionally discovers in particularly pretentious wine lists the lingering echo of this classification, with 'empyreumatic' applied to a smokey red wine. But it still seems an inadequate list to capture such common smells as fruit or salami or flour or gin or dust.

A more subtle attempt was made in 1916 by the German psychologist Hans Henning who chose six primary attributes of smells, and presented them at the six corners of an imaginary prism. Every smell could, he hoped, be located somewhere within the odour prism, its location showing how close or distant it was from each of the six primary attributes. His choice of six were: flowery, putrid, fruity, spicy, burnt, resinous. The attempt failed, for two obvious reasons. One reason is that smell is very personal, and different people locate smells within Henning's prism in markedly different places. The second reason is that there are plenty of smells which do not relate easily to Henning's choice of six primary smells. Henning's attempt had looked promising for a time, but it proved to be yet another expedition into the hinterland of smells that failed, leaving just a few abandoned prisms behind.

In the 1920s a pair of American scientists, Crocker and Henderson, attempted a more technologically sophisticated approach. They used four characteristics (fragrant, acid, burnt and 'caprylic' – goaty) with a scale of 0 to 8 for each. Any smell was classified with four numbers, one for each characteristic. So a rose was given the number 6423, which meant that it was strongly fragrant, had some acidity, and a little less of the other two characters. Vanilla was numbered 7122 (very fragrant, hardly acid at all, and very slightly burnt and caprylic). The smell of ethyl alcohol was given the number 5414. Crocker and Henderson worked out the numbers for several hundred smells by trial and error, relying on their personal perceptions, not on any machinery. It is said that Crocker would ring Henderson of an evening and call out a number – 6443 or 8257 – so that Henderson could guess what substance it referred to. Having perfected the numbers, they marketed an olfactory kit which contained 32 glass vials of varied smells for sniffing, and a manual giving the numbers developed by Crocker and Henderson. The kit was sold to oil companies, soap manufacturers, the army and the police, but ultimately it proved to be yet another addition to the pile of rusty junk from the failed expeditions into the jungle of smells.

Almost every year another attempt at listing the primary smells is made. Time after time the attempt is hailed as a breakthrough, and then forgotten. Why have all these attempts to find the primary smells, and the single system for relating all smells together, failed? First, these attempts to find the basic smells are defeated by the disorderly nature of smells. Simply put, there are no primary smells. There are huge numbers of odour-carrying molecules that come at us in a huge variety of combinations. There is no natural spectrum of smells, in the way that there is a natural spectrum of colours. In addition, there is no hierarchy of smells. There are a few, but only a very few scents that the brain recognises as generic, but they apply only to a minority of the smells we encounter in ordinary life. Otherwise smells come at us in a disorderly cloud of airborne molecules with no hierarchy.

There is another fundamental problem with trying to establish simplified systems for categorising smells – they are for the most part not simple nor are they easily ascribed to one category. Many are multi-dimensional, with significant touches of several classes. The rich scent of leather, for instance, is compounded of the animal skin, the oak-based tannin with which leather is cured, and touches of fruitiness and fishiness.

Finally, we must recognise that the mind is not interested primarily in the smell, but in the thing which gives off smells. Take, for example, a Canteloupe melon ripening in a corner of the kitchen. We smell a complex mix of molecules given off by the melon which (with an effort) we can analyse, but we say to ourselves 'I smell a melon'. If we used Zwaardemaker's categories, we might say that it gives off a vigorous ethereal volatility that is mainly fragrant, but which has a petrolly aromatic smell, and a touch of hircine cheese. If we used the Crocker/Henderson kit, we might describe it by the numbers 7443. But, self-evidently, neither approach works. We smell a ripe fruit, not a fragrant, ethereal, aromatic and slightly hircine smell. It is the *thing* that we want to recognise, not the smell, and the smell is only a clue to the thing.

To look for a vocabulary of smells that starts from a simplified classification is to go down a blind alley. It does not satisfy our need to label smells by association with the things that give off smell. It does not satisfy our recognition that smells have several dimensions, and that one dimension may be fragrant in a floral way, another may be fragrant in a grassy way, and yet another may be acid in a fruity way, or smokey in a spicy way. It does not acknowledge the fact that there are dozens of volatile molecules in any smell, and our brains recognise the pattern of molecules, not the individual molecules that make the pattern.

Perfumers and aromatherapists describe smells in quite another way. They have deliberately adopted a musical metaphor for describing perfumes, in recognition of the fact that smells have depth and complexity. So perfumes for them have three sets of notes, harmonising together to make what perfumers call the scent's '*accord*' (French for a musical harmony). The notes express themselves over time as the ingredients evaporate, with the immediate impression of the top notes (also called head notes) leading to the deeper middle notes, and the base notes gradually appearing in the final stage, sometimes described as the 'drydown'. A perfumer creates this accord carefully, using his knowledge of the evaporation process of the various components of the perfume.

Top notes consist of light, small molecules that evaporate quickly, and so are perceived immediately. Since they form the first impression of a scent, they are very important in selling a perfume. They might be described as light, fresh, green, citrus, assertive, sharp, crisp, refreshing. Middle notes (sometimes called heart notes) emerge just before the top notes dissipate. They form the body of the perfume, and often mask the initially unpleasant impressions of base notes. Words that might be applied to these middle layers in the perfume are mellow, rounded, rich floral, spicy, fruity. Base notes (or background) bring depth and solidity to a perfume. They are typically rich and deep, and are usually not perceived until half an hour after application.

When describing a perfume in the round, rather than by its many components, perfumers use a traditional classification which dates back to around 1900, but they recognise that many perfumes contain aspects of different families. So a perfume may be described as 'a single floral' dominated by the scent of one particular flower; 'a floral bouquet' blending several flower scents; 'an oriental' perfume featuring the sweet slightly animal scents of ambergris or labdanum, often combined with vanilla, tonka bean, flowers and woods; 'leather' featuring scents of honey, tobacco, wood and wood tars; '*chypre*' (meaning Cyprus in French) built on an accord consisting of bergamot, oakmoss and labdanum; '*fougère*' (meaning fern in French) built on a base of lavender, coumarin and oakmoss. More recently, new families of perfumes have been developed, such as 'green' perfumes with scents evoking cut grass, crushed leaf and cucumber; 'oceanic' perfumes giving the clean smell of the sea; 'fruity', featuring the scents of fruits other than citrus such as peach, mango or blackcurrant; and 'gourmand' perfumes, with components designed to resemble food flavours.

A convenient way of grouping perfumes, much used by sales people, is the fragrance wheel which was created in 1983 to simplify the identification of fragrances.

It is tempting to think that perfumers have cracked the problem of naming smells. They have certainly found a successful metaphor for describing the complexity of smells based on musical harmony. But nature is not a professional perfumer, and there are many smells that come at us in a disorderly way, with top notes that are more like bass notes, and bass notes that are more like top notes. The very precision of perfumers' language makes it difficult to apply in the ordinary world, where we simply do not have time and mental capacity to analyse a smell for 30 minutes to see how it evolves over time. More usually we have four or five breaths before we cease to register that smell and find a different one swimming into our ken. The language of the perfume trade is too dependent on expert knowledge of ingredients that few have come across in real life – how many people know what oakmoss or labdanum or musk or ambergris or ferns smell like? (Oakmoss is a lichen that grows on Indian oak trees; labdanum is a resin of cistus plants; musk is an internal organ of small deer; while ambergris is an internal deposit in whales.)

Perhaps the most serious problem with the language of perfumes is that many people who are not perfumers find it faintly suspect. It belongs to a luxury world which is far from down to earth, one where willowy men and women in sharp suits talk down to the humble purchaser. Just as the men I met in the Surrey hardware shop thought, it seems like a joke to apply the language of perfumes to ordinary things like wellington boots or linseed oil.

In their search for reliable ingredients for perfumes and fragrances, perfumers have spawned a valuable industry in the production of synthetic odorants. As a result, we now can name a large number of molecules that are recognised by the human brain as ingredients of smells. There is, therefore, a language of smells that is highly precise. However, this is a language that serves the purpose of classification, not of communication. If, God forbid, I wanted to convey to someone the smell of bad breath, I would seek words that conveyed its putrid nature, its affinity with rotting flesh or overripe cheese. It would be open to me to use the chemists' language of molecules, to say 'the smell is tetramethylenediamine and pentamethylenediamine', and I suppose it might impress. I might use the slightly more down-to-earth synonyms for those molecules – putrescine and cadaverine. But it would not be an effective way of communicating to someone unversed in the chemistry of odorants so that they could get a mental picture of the smell. It would only close down the communication.

There are more obscure corners of the olfactory jungle that have been explored. Here is a list of the smells of mushrooms. I admire it for its dedication to sniffing and its simplicity of language:

> **Mushrooms**
> I take a piece of the mushroom (or a whole cap in the case of small mushrooms) and crush it between my finger and thumb before trying to assess an odour …
>
> 'Not distinctive' is probably the most common mushroom odour, but distinctive smells include:

- Farinaceous or mealy. Often compared to the odour of cucumbers, watermelon rind, or an old grain mill. Common in many mushrooms, including *Polyporus squamosus*, *Agrocybe praecox*, *Mycena galericulata*, *Tricholoma sejunctum*, *Clitopilus prunulus*, and *Entoloma abortivum*. Some mycologists subdivide 'farinaceous' into three odour groups: strictly farinaceous, cucumber/farinaceous, and rancid-oily-fishy/farinaceous.
- Foetid-*Russula* odour. Often compared to benzaldehyde; to me it smells like maraschino cherries that have gone slightly bad.
- Fishy or shrimplike. Examples include *Lactarius volemus* and *Russula xerampelina*.
- Spermatic. Primarily in species of *Inocybe*.
- Like anise (the flavouring in ouzo or black liquorice). Examples include *Clitocybe odora* and *Agaricus*.
- Like green corn. Examples include species of *Inocybe*.
- Like bleach. Primarily in species of *Mycena*.
- Like swamp gas or coal tar. Primarily in species of *Tricholoma*.
- Like apricots. Primarily in species of *Cantharellus*.
- Like almonds. Primarily in species of *Agaraicus*.
- Like garlic. Primarily in species of *Marasmius*.
- Phenolic. Primarily in species *Agaricus*.
- Foul. Any mushroom can smell foul after it has begun to decay, but some have a strongly unpleasant odour anyway. Examples include Stinkhorns and *Lepiota cristata*.

(extract from M Kuo at Mushroomexpert.com)

Perhaps the most successful approaches to the naming of aromas have been developed by the drinks trade – notably for describing wine, whisky, beer and coffee. They tend to start from the right place – the tricky business of applying one's nose to a glass before moving on to describing the aromas in a fairly systematic way. For the whisky connoisseur, the business of sniffing is called 'nosing'. Expert tasters add a little water to the whisky supposedly to release the fragrance and to avoid anaesthetising the nose. They recommend letting the whisky 'breathe': that is, letting it sit in the glass undisturbed for a time, before swirling round the glass and nosing it. Wine connoisseurs talk of the three '*coups de nez*' – the three nosings. The first coup consists in putting your nose to the glass without disturbing it, to perceive the most diffusible elements of aroma that it gives off. The second coup is a light swirl to increase the surface area of wine in contact with the air and so encourage the evaporation of aromatic molecules. The third coup is for the harsh critic – the glass of wine is agitated in a brisk manner so as to break up the wine and release the least volatile molecules. This method tends to bring out the less agreeable smells such as sulphur, rotten wood or mould.

Thus primed by swirling, sniffing and nosing, the connoisseur turns to the words to describe what he smells in his drink. The wine trade now has a systematic approach to describing the aromas of wine, based on the work of a US chemist called Ann Noble. Her aroma wheel, developed in 1984, simply lists the smells that one may perceive in a glass of wine, so that wine buffs can be prompted to identify with precision what they experience. The smells are grouped in a circle, in recognition that there is no hierarchy of smells, with related smells placed close together. So the wheel gives 90 names from which a drinker can choose the names that fit best, including honey, butterscotch, soy sauce, wet cardboard, sulphur dioxide, soapy, fishy, lactic acid, sauerkraut, sweaty, mousey, orange blossom, liquorice and so on.

The wine aroma wheel has spawned a garage of other wheels. You will find a beer aroma wheel, a whisky aroma wheel, a coffee aroma wheel, a cigar flavour wheel and so on. These are just lists of words that occur to people sniffing these drinks, arranged in a circle. They are not a complete vocabulary of smells. But they do put into order the most likely descriptors that people use. The coffee wheel, for instance, has nine main categories – the aromas that are inherent in the plant: flowery, fruity and herby; the sugar-browning aromas that are created by thermal reactions during roasting: nutty, caramel and chocolatey; and the dry distillation aromas related to the burning of plant fibres during roasting: resinous, spicy and carbon. Alongside the coffee flavour wheel there is a smaller cog – the faults wheel for such problems as mould, underripeness or taint.

These are practical approaches, down to earth and, best of all, they are used. If the aroma wheel could be expanded to include all the smells of everyday life, it would be the answer we need to naming smells. But with 90 names just for a glass of wine, the aroma wheel would be as big as a Ferris wheel to accommodate the smells of a farm, or a dusty attic, or a suburban street, or a wood. It would be just too cumbersome.

We are looking for a different approach to the naming of the whole range of smells in ordinary life, not primary smells which have proved to be mythical, not lumbering great wheels. We need something that is practical, open to everyone – not just experts – and communicative.

CHAPTER 13

On the Tip of My Nose

Many people believe that smells are beyond words. Some think that we have no vocabulary for smells. Some think that the sense of smell sidesteps language in our brains. Some think that only specially-talented people like perfumers or wine buffs can put words to smells. These are myths that have not been properly challenged.

One of the leading researchers into how smells are translated into words is Jonas Olofsson of Stockholm University. He calls smell 'the muted sense'.[26] It is the sense that struggles to find the right words. It is not, however, a silent sense. Olofsson says that our difficulty putting smells into words is the cumulative effect of problems at three successive stages along the pathway from smell to language. First, we have difficulty perceiving the detail in a smell; we perceive it as a single composite 'chord', and so we do not get the wide range of clues that we get when we look at an object. We can see that an object has a yellow skin, is slightly curved, and comes in a bunch; but all we get by smell is the integrated banana aroma. Second, because our brains quickly endow a smell with an emotional value, we do not have many varied ways of building up the representation of the object in our heads. With sight, our brains can size up an object from several angles before putting a name to it. Third, the business of verbalising requires our brains to play around with top-down and bottom-up searches in our mental dictionaries: top-down approaches which guess at a likely label, and bottom-up approaches which use detailed cues to check whether the label fits. That works well with vision, and badly with smell, which often leaves us guessing, and guessing wrong.

I decided to challenge the myths in the simplest way, by taking some objects and seeing whether I could jump some of the obstacles to putting their smells into words. I took an old bottle from the back of the drinks cupboard, a small bottle with an oversized label, that looked like a small schoolboy wearing an oversized school uniform for the first time. It was a bottle of Angostura bitters, the aromatic pinky-brown ingredient used in Manhattan cocktails, and in pink gins favoured by Royal Navy officers. The label, which sported an engraving of the medal that the drink won in 1873 at some international exhibition in Vienna, was stained; the neck was crusted with sugary residues; this bottle had not been touched for years. I must have inherited it from my father when he died.

Angostura bitters comes from Trinidad and Tobago. It is a botanically-infused, extremely concentrated alcohol mixture. The recipe was developed by Dr J.G.B. Siegert, surgeon general to Simon Bolivar's liberation army in Venezuela, who had his base in the town of Angostura. He first used it as a tonic for soldiers with upset stomachs, but went on to set up a distillery as a commercial venture in 1830. Local Amerindians are said to have contributed little-known plants to the recipe which remains a closely-guarded secret, known to only five people at any one time.

I had spent a happy evening in my teenage years experimenting with cocktails, with Angostura bitters as one ingredient. I remember trying to reconstruct Pimm's No 7 and failing to get it to the desired level of alcohol (not for want of some lavish tasting). Since then I had not touched the bottle, but I have picked it as my talisman for the naming of smells.

Now I take the bottle up for the first time in years. I unscrew the crusted top, and pause. What will be the best way to describe the fragrance that the bottle gives off? There are some wrong answers. It will not do to describe it as the unmistakable scent of Royal Navy wardrooms, of pink gins, of Venezuela. Those may be exotic descriptions, but they convey nothing to people unacquainted with the Royal Navy, pink gins or Venezuela. I could describe it as a velvet, crimson smell, with flashes of gold and purple, reverberating like a chord from a grand symphony. Too poetic, too synasthaestic. I might say it is the smell of botanical bitters and concentrated alcohol – bitter, botanical, alcoholic and aromatic. That is just a tautology and it adds no fresh information. Or I might describe it as the smell of teenage experiments with cocktails and my father's drinks cupboard. Too personal.

So let us see what words will work. I take a first sniff. It is easy to recognise so distinctive a smell, despite the lapse of time since I last encountered it. But if I simply opt for recognising the well-remembered aroma of Angostura bitters I will not be able to analyse its components. So I ask myself an oblique question – is this a rich, complex aroma with many layers, or is it a simple one? It is indeed a very rich smell, and so complex that I know straight off that I am dealing with a blend of flowery, fruity, fermented and alcoholic smells. Now a difficult choice has to be made: which is the dominant smell? I already know from the first sniff that there are competing aromas to be considered. I take a second sniff – it is not clear – there is both a dark treacly sugar aroma and a spice smell fighting for dominance. Maybe it does not matter which is the principal aroma; maybe it is good enough to settle for spice and treacle. With my next sniff I hope to identify some of the deeper layers. I already know there are several. There is a fruit aroma that is surely citrus, maybe grapefruit or orange. But I will need another sniff and some real concentration to get at the other layers. Another sniff tells me that there is a warm spice, of the kind I might find in mulled wine or Christmas cake. There is still a floral fragrance to identify, but I cannot place it. Finally I sniff to check what I have missed – yes, there is a telltale piquancy of alcohol – it is faintly irritating to the nose but not like the prickle of pepper or smoke.

So my sniffs have told me that, though smells may at first seem to be inarticulate, once they are broken down they can be coaxed into speech. There are angles, attributes to a smell that can be interrogated and which give a basis for describing it in ways that are communicative to other people.

- *First*, a smell can be described as complex with many layers, or simple. Complex, rich smells have layers that can, with a bit of concentration be unpicked and described. The simpler smells (I think of rubber, grass or ammonia as simpler smells) are much harder to describe, but thankfully most of the smells of the natural world have some complexity.
- *Second*, we can describe which part of the natural world a smell belongs to, be it fruity or malty or meaty or smokey or floral or chemical. And that applies to each of the discernible layers in a complex smell. It becomes particularly descriptive when we encounter a smell that has attributes from the wrong part of the natural world – those qualities that do not seem to belong to the same part of the world as the main substance. I think for instance of the smokiness of demerara sugar, or the berry smell of some leather or the sweaty smell that lingers after you sniff a banana.
- *Third*, we can describe a smell in terms of other smells that it evokes. It might seem to be a variant of a neighbouring smell, in the way that grapefruit, lemon and orange are close neighbours and can be described in relation to each other. A smell might seem to stand alone but have affinities with other unrelated aromas, in the way that some plants smell of cat. Or a smell might seem to stand at an extreme: the smokiest or richest or sharpest of its kind.

- *Fourth*, we can describe where an odour stands on the bitter/sweet spectrum. This is never simple. On our tongues bitter and sweet are simple tastes. In our noses bitter and sweet are nuanced. Creamy sweetness as in vanilla; resiny sweetness in cedarwood; grassy sweetness in the breath of cows chewing the cud; heavy sweetness in hot tar; religious sweetness in incense; artificial sweetness in strawberry-flavoured food. Tart bitterness in soft fruit; zesty as in lemon; sour as in milk or cheese; acrid as in smoke; sharp as in animal urine. Furthermore, while our tongues make a sharp contrast between bitter and sweet, in our noses these qualities are intertwined. Many fruits are bitter-sweet, and the quality of bitter-sweetness is just as nuanced as the quality of bitterness or of sweetness. Grapefruit is bitter-sweet although the emphasis is on sweetness; vinegar is bitter-sweet with the emphasis on sourness; brandy is sweet with an edge of burnt bitterness; sharp apples (like Granny Smiths) hover over the midpoint between bitterness and sweetness.

- *Fifth*, many smells have a piquancy, a tendency to 'get up our noses' that can be put into words. The eye-watering quality of onions; the punch of ammonia; the sneeze provoked by pepper; the tingle of lemon; the sweet piquancy of ginger – all these can be articulated.

These five qualities – complexity, family resemblances, affinities to other smells, bitter-sweetness, piquancy – are probably enough to get along with. They are not, however, the end of the story. A good description of some smells might say something about how they evolve over time, from the first whiff to the lingering aftersmell (which perfumers call 'the drydown'). It might identify some attributes like how old or fresh a smell seems. It might say something about smells that are gentle at a distance and overpowering close up. It might refer to the feel of a smell, be it velvety or gummy.

All this can be done without fancy words: no need for empyreumatic, or mephitic, or musky. All this can be done without extravagance: no need for the exotic or the erotic, or the mystique of the Orient. All this can be done without invoking the names of molecules or colours or abstruse analogies to music. It can be done without inventing new words like 'brunky' or 'brambish'. There is, in short, a simple everyday language for smells at our fingertips, and it can be built on. At the end of this book are descriptions of over 200 everyday smells. My hope is that others can improve on these descriptions and so, slowly, by accretion, a rough consensus can be built up which will allow people to describe smells to each other and to themselves. To be able to compare notes amongst each other about what things smell like is an achievable ambition, but I have a less achievable ambition as well – to have descriptions of aromas that are so vivid and accurate that the description alone would evoke the smell without any molecules passing our noses. Then, in my ideal world, I could lie back, recite a description of the scent of, say, fruit cake or new-baked bread or a lavender walk, and summon up the aroma in my head. But that is probably an ambition that can never be realised.

It is of course no easy matter to put smells into words, but the problem is not a shortage of words: there are plenty of them. The problem lies elsewhere. I note in passing that the ease we have in describing colours is not because we have a vast vocabulary of specialist terms for colours. The primary colours have had simple words for millennia, but some of the more nuanced words for colours are of quite recent invention. Mauve, for instance, was first introduced into the language in 1859, when aniline dye was first derived from coal tar. Viridian dates from 1882. Chrome yellow dates from 1797; cadmium red from 1822; postbox red from 1859. Cyan, the purple-blue which is one of the three fundamental hues of colour printing, dates from 1879. More recently, the advent of a vast spectrum of coloured paints has given us fresh words for colours like avocado or celadon or willow green. There are those who think that smells could be described better if we had a vocabulary of words that applied only to smells. There are some such words, but they are not useful for everyday communication. *Rancio*, as we saw above, is a vintner's word for the particular bouquet of heavy fortified wines stored in oak, and it conveys a buttery, nutty aroma with an edge of tannin. Feinty is a whisky blender's term for the leathery smell of the liquid drawn from distilled barley in the later stages of distillation. In the world of perfumery, the terms for animal smells such as musk, ambergris and civet refer to odours that most of us have not encountered directly. We can describe smells perfectly well without these specialist terms.

The problem is not the words; the problem is the fleeting moment of perception that our noses and brains allow us. We breathe in, then out; the next breath brings us a new set of molecules to sniff. Because the time is so short to analyse the odour, we tend to rush to recognise what the smell is. Having recognised it (or more often having failed to recognise it) we cannot then pick it apart in order to describe it. There is a simple solution to this problem. It is a series of short, sharp sniffs. Each sniff should have a purpose, a question that needs to be answered. The first sniff – is it a simple or complex smell and which family of things does it belong to? Second sniff – what other smells is it like and what distinguishes it? Third sniff – what are the lower layers like? Fourth sniff – what quality of sweet, bitter or piquant have I missed? There may be a need for more sniffs, because smells are elusive and we will not capture them first time round.

There is no special art to a short sharp sniff. We do the right thing naturally. An Australian psychologist, D.G. Laing, ran a series of tests in the 1980s to establish how people sniff and how their different approaches affect their ability to detect smells and to assess the intensity of smells. His tests showed that people vary markedly in their patterns of sniffing. They have characteristic patterns, as distinctive as their fingerprint, which they maintain with different tasks and different odours. Some are short sniffers; some are long sniffers; many are Morse code sniffers in patterns of long and short. What is the best style of sniffing? The answer that these tests threw up was that it is very difficult to improve on the efficiency of people's ordinary sniffing techniques, and that a single natural sniff provides as much information about the presence and intensity of an odour as does a series of many sniffs. Even if we prolong a sniff, or if we do a series of sniffs, the work is done in a fraction of a second that does the work. Only when we are dealing with a complex mix of odours does a series of sniffs help to discriminate between the components. But, faced with a complex mix, there are limits to what we can discern. Even the most experienced sniffer cannot identify more than three or four components in a mix. This limit cannot be increased by training – it appears to be imposed by physiological constraints in our noses or processing constraints in our brains.

I take up my bottle of Angostura bitters again. Now I know that there is a simple technique to sniffing, I can be more systematic. The technique is the short, sharp sniff; a pause to clear the head; and another short sharp sniff, repeated until I can identify three or four components in the many-layered mix. I open the bottle, and take a first sniff for complexity. It is as I thought, a rich cake of layers, some fruity, some spicy, some treacly, all concentrated, and with an edge of something that is almost medicinal. Perfumers deliberately clear their heads between sniffs by smelling something quite different, often using coffee beans as their means of resetting their noses for the next sniff. I clear my head by smelling a jar of instant coffee. The second sniff is for the main resemblances: I detect a warm spice like ginger, and citrus, somewhere close to grapefruit and orange. The third sniff is for the lower layers. I can detect a flower that is almost bitter, a scent I have encountered in some continental drinks like Suze, and I think there are more spices in the mix than just ginger, maybe nutmeg. The fourth sniff is to check what I have missed: yes, I have missed the alcoholic headiness.

So my sniffs have given me this description of the smell of Angostura bitters – 'a rich, multi-layered smell, dominated by warm mixed spices and treacle, with a strong element of sweet citrus and an unusual floral scent, and with a faded alcoholic headiness'. Now I look at the heavily-stained label on the bottle, which reveals that the bitters are 'a skilfully blended aromatic preparation of gentian (which must have been the difficult-to-place floral smell) in combination with a variety of harmless vegetable spices and vegetable colouring matter. 78.5 per cent proof'. My own description was not too wide of the mark.

Here, then, is my approach to naming smells which anyone can use. It is an approach which, like so many things, will improve with practice. It requires four purposeful sniffs. Four short, sharp sniffs, with a pause in between to clear the head. Sniff One: is this a complex or simple smell? Sniff Two: what other smells does it seem like? Sniff Three: the lower layers. Sniff Four: what have I missed?

Once we have overcome the thought that smells are too difficult, we can begin to read the meanings of smells. The thing to realise is that our noses are tuned in to the world of molecules. It is a world that is invisible to the naked eye, invisible even to the conventional microscope because molecules are smaller than the wavelength of light. It is a world that is inaudible. But we can smell it. Our noses, which seem so ineffective, are detecting tiny concentrations of tiny molecules; no machine has yet been invented which can do the job better than our noses.

For organic chemists this language is telling a story about hundreds and thousands of complex molecules with incomprehensible names. For those of us who are not organic chemists the language of smells is telling a more intelligible story about processes of chemical change, processes such as fermentation, searing, bacterial decay or fire. When we catch a whiff of a compost heap, it has the richness of fruit cake or of ripe cheese. The smells of all three are the product of a process of bacterial decomposition that they all share. When we catch a sniff of sulphur from a burning match, it has a tang of the seashore and of the smell of our urine after eating asparagus. The smells of all three have a constituent in common. When we smell the delightful warmth of grilled steak, it has something in common with the smell of beer and the London Underground, because all three involve the same chemical reaction. When we catch a hint of smoke in whisky or brown sugar or rosemary, that is because they have molecules in common with wood smoke. There are some common features to smells, even if there are thousands of varieties of smelly molecules and millions of combinations of molecules.

There are interesting chemical stories behind many of the ordinary aromas we encounter every day. Let us look, for example, at freshly-baked bread, which, according to one of the big-circulation tabloid newspapers, is the nation's favourite smell. Like most things in nature this is a smell made from a mix of hundreds of different types of molecule. But the big contributions to the smell of new-baked bread come from three chemical processes. The most important of these is the process that gives us the browning of the crust and its delicious smell. This is the reaction of amino acids and sugars in heat, and it is known as the Maillard reaction after the French chemist who first described it in 1912. Seared steaks, browned onions, toast, chocolate and roasted coffee all share this reaction. Their aromas have something in common with each other, and with new-baked bread, because some of the molecules they share are the result of Maillard reactions. Maillard reactions occur quickly in cooking, and they can also unfold more slowly in food stored at room temperatures. The scents of ripening cheese, for example, are due in part to the Maillard reaction. It is also the secret of Serrano ham, a raw meat that is dry-cured for several months.

There is another process at work in bread as it is baked: it is the caramelisation of sugars which gives a sweetness to the aroma of bread. Unlike the Maillard reaction which involves the effect of heat on amino acids, this involves only the effect of heat on natural sugars. The same process is found in the smell of roasted malt. All this is in the smell of the crust; inside the loaf, in the crumb, there are other smells with connections to other substances. One little molecule that is part of the aroma of bread, called (E)-2-nonenal, is also a component of the smell of cucumber. And indeed, if you approach a cucumber with an open mind, not predisposed to smell only fresh and green aromas, you will get a whiff of bread from it. That explains the unexpected pleasure of a cucumber sandwich. This same molecule is also linked to changes in human body odour as we age; it is part of the dull sweet smell of old people and their houses. Grandparents, cucumbers and bread are in this way connected by links that only our noses can detect.

A whiff of new-baked bread is more than a simple pleasure. It is full of meanings about the inner nature of bread, and the hidden connections between bread and other things. Let us take a different object to illustrate another important smell-producing process. Salami is my chosen object, and the process it illustrates is fermentation. Salami, pickles, yogurt, olives, cheese, sauerkraut, shrimp paste, these are all the results of fermentation. These all smell different, but they have in common a power and richness of aroma which is the sign that they are the result of a common process.

A few slices of Italian salami give off a complicated stink of pork and (sometimes) garlic, rich and old almost like leather, with many layers which include a sweet fruitiness, a cheesy quality, a hint of the medicinal and a hint of smokiness. The quality of smell is more, however, than just the smell of the ingredients – pork and beef, or frequently horse or donkey meat, with some white wine, garlic and pepper. Those ingredients have been mixed together and stuffed into an edible casing, hung up to dry slowly, and left to ferment. The sausage casing effectively excludes oxygen, and allows fermentation in which a series of chemical changes happen in the meat, and the characteristic aroma and flavour of salami slowly emerges.

People have been fermenting food and drink for millennia. The earliest evidence of an alcoholic beverage, produced through the fermenting of rice, fruit and honey, dates from 7,000 BC and comes from a neolithic village in China. There is plentiful evidence of controlled fermentation from ancient Babylon, from the Egypt of the Pharaohs, from South America before colonisation and from central Asia. As well as producing alcohol, fermentation serves to preserve food and drink, to develop a richer range of food components like proteins and vitamins, to cut down cooking time, and to make a richer variety of flavours.

Put through a gas chromatograph, the aroma of our slices of Italian salami is found to be composed of hundreds of different molecules, of which 80 or 90 give off discernible odours. Among the most important odorous molecules are many that are found in other fermented foods, and that connect salami with quite different substances. So there is 1-octen-3-ol, a vegetable broth smell found also in baked potatoes; guaiacol, which is an important part of the smell of wood smoke; 1-octen-3-one, given off by mould and mushrooms; 2-and3-methylbutanoic acids, given off by cheese and old socks; a range of fruity molecules; and diallyl disulphide which comes also from garlic. Here again is the hidden language of smells: the secret code that connects one substance with another, that only our noses can discern.

The world as it is perceived by our noses has a different weave to the world as we perceive it through vision or hearing.

Another strand in this weave is the product of sulphur. Sulphur is the smell of rotten eggs, and, in the popular imagination, of the devil. Our noses sense a sulphur-laden molecule instantly, and it has an edge, whether it comes from a safety match or an onion or an oil refinery. The object I have chosen to illustrate the sulphur strand is a plate of asparagus. This delicious vegetable has a remarkable and potent impact on the scent of urine. Asparagus uniquely (as far as we know) contains a sulphur-laden acid called, unenterprisingly, asparagusic acid. This is not volatile at the point when we eat it, so it gives no smell. But as we digest it, it breaks down into molecules that are volatile, and straight away taint our urine with the dodgy stink of sulphur. Marcel Proust, the great French writer who wrote brilliantly about all the senses but particularly about the sense of smell, appreciated the smell of asparagus, and wrote that it 'transforms my chamber pot into a flask of perfume'. But most of us find it a suspect smell, and its edge of sulphur connects it with some more ferocious stinks.

When it comes to the worst stinks, sulphur is usually the vital ingredient. At the less extreme end of the spectrum, relatively innocent stink bombs are made of ammonium sulphide. At the most extreme end, the US government and military have experimented with a product called 'Who Me?' as a chemical weapon for controlling riots and for flushing enemy soldiers out. It is a mixture of five sulphur-containing chemicals, and smells like rotting carcasses. It has not yet made it to the battlefield, as the problem has proved to be how to deliver it against an enemy without affecting friendly soldiers. Sulphur is not always a stinker: in mild doses it gives to some things, like blackcurrant leaves or the stems of tomatoes, an edge which is stimulating without being disagreeable.

Asparagus has thrown up a curious side issue. Something like one in five people do not notice any post-asparagus smell in their urine. A study in 2010[27] of 10,000 people analysed this peculiarity, and checked the genetic sequencing of the people who could not perceive the smell of asparagus, to determine whether they did not produce the sulphurous molecules at all or simply did not detect the smell. It seems that the defect is with detecting the smell, and that it stems from a minor genetic mutation among the cluster of genes that code for smell receptors. Here is another indication of the variability of this sense from person to person.

Another shared feature of many aromas is not really a product of the sense of smell at all. It is piquancy: the attribute of some smells that gets up your nose, make your eyes water or takes your breath away. The sensation of piquancy is transmitted to us by nerves that register feeling – the somatosensory nerves. Because piquancy comes to us as a part of a mix of smelly molecules, we perceive it as being part of the smell, even though it is transmitted by nerves that are not olfactory nerves. And because piquancy is accompanied by a variety of smells, it has a variety of characters – the sweet piquancy of ginger; the tickliness of pepper; the tart zestiness of lemon; the sharp spiciness of mustard – all these are varieties of piquancy. Equally the breathtaking quality of ammonia; the hiccup-inducing shock of chilli; the gasping, eye-watering sensation of smoke: these too are variations on the same theme. Scientists distinguish between milder heat and stronger heat, calling the first piquancy, and the stronger sensation pungency. Mustard is piquant; chilli peppers are pungent. At the fiery end of the range, there is an official scale for assessing the strength of the sensation, the Scoville scale, on which pimento scores a mere 100 Scoville heat units, while the hottest pepper of all (Carolina Reaper) scores over a million. We less precise people have a wide range of words for describing piquancy or pungency, according to the smells that it is mixed with – hot, tart, zesty, sharp, acrid and so on.

Smoke deserves a special mention, not only for its acrid quality, but also for its presence in many smells, including smells that are not the result of fire. It is perhaps the feature of the olfactory world that is most easily recognised and that carries the most obvious meaning. Human beings are extremely sensitive to the slightest whiff of smoke. Small wonder, as we are the only creature that can unleash fire: we are both its masters and its potential victims. It is a matter of immediate importance to us whether the smoke we detect is safe or menacing, whether it comes from a controlled fire or one that might put us in danger. It is so important to us that we have, from ancient times, made smoke a feature of our religions. Incense smoke rises to the heavens; it carries the souls of people towards the gods; it pleases the gods with the smell of sacrifice; it makes our prayers visible; it exalts us.

A good bonfire is one of the richest, most complex and most evocative smells. It is the smell of autumn, of woodland being cared for, of gardens being tidied, of ruminative old men watching the slow smouldering and sudden outbursts of flame and thinking of the dissolution of life, of children at fireworks night excited by an ancient force. It is a smell that evolves over time as the bonfire changes from the first vaporous billows of smoke to the dying embers and damp ashes. It is a smell that varies from the gentle sweetness of a distant whiff to the acrid power of the smoke close up.

Bonfire smoke has many constituents, and well over 400 different molecules: plenty of solid particles of ash and half-burnt leaf whisked up by the flames; plenty of water vapour produced as the wet material is burned; plenty of sulphur and carbon; plenty of hydrocarbons yielding a wide palette of other compounds. Chief amongst these last are the aromatic alcohols that are the telltale signature of smoke, to be found even in substances that have been nowhere near fire. Guaiacol and syringol in particular are the products of burning wood – more guaiacol in the smoke of hardwoods; more syringol in the smoke of pine and softwoods. These and other molecules are what give smokiness to substances that have no connection with fire, for instance to wines that have aged. They account for the aroma of oak smoke in many mature red wines, and the aroma of gunflint in some white wines. They are what make the scent of incense given off by some plants like cistus and rosemary. They are what give the hint of smokiness to such diverse things as leather and unrefined sugar. And of course they are present in the aroma and flavour of smoked foods and of drinks based on barley malt smoked over wood or peat. The underlying smokiness of many aromas is yet another hidden meaning to be found in the language of smells.

Still in the open air, we can find another repeated motif of smells – the smell of bacteria at work. It is everywhere in the earth, and it is to be found in concentrated form in the good stench of a compost heap. I happen to be a fan of compost. More than a fan, I see myself as the father to a small family of compost heaps which demand frequent love and attention, and which repay me with valuable humus and a fine smell. The smell of compost heap is not, I acknowledge, a pleasure for everyone. For many it stinks of foulness, decay, of the products of the gut and the life of moulds and fungi. And the secret of a compost heap is indeed its vast population of minute creatures that consume plant material, and the creatures that prey on them. So a compost heap is full of bacteria that break down woody material; and fungi, moulds and yeast that digest the material that the bacteria cannot digest; and earthworms digesting partly composted material; and rotifers (tiny, maggoty inhabitants of moist soil) which eat decomposing material. It is these populations of largely invisible creatures which speed the transformation of the compost heap and create the smell of vegetable decay.

This smell of bacterial activity is a common sign that nature is at work. All cheeses, for instance, depend on bacteria for their very existence, but one bacterium, brevibacterium linens, is responsible for the real stink of the highest cheeses. Brevibacterium linens requires a warm, moist environment in order to flourish. It lives chiefly on the surface of a cheese, and the most fragrant cheeses like Epoisses, Limburger, Stinking Bishop, or Gorgonzola Piccante are washed regularly in juice or brine to encourage the bacterial growth. Stinking Bishop, for instance, is washed in pear cider made from the Stinking Bishop variety of pear. But the smell of bacteria is to be found in many other places as well as cheeses – in leaf mould under trees, in animal dung and in our own guts.

So the meanings of smells can be slowly opened up to our minds, if we are receptive. Once we can learn to unpick a smell, to identify its layers and its affinities to other related odours, once we can put this, however haltingly, into words, the meanings begin to emerge. There are meanings of several different kinds. Personal meanings tell us of our own childhood past – the smell of our grandparents' houses or our first school playground. Factual meanings tell us what substance is responsible for the smell, be it new-baked bread or a compost heap. Chemical meanings tell us about the processes that a smell has undergone, such as the Maillard reaction or the work of bacteria. There are also social meanings, and I set out next to explore those social aspects of smell in the heart of London.

CHAPTER 14

The Most Expensive Perfume in the World

'We are divided from our fellows by our sense of smell. The working man is our master, inclined to rule us with an iron hand, but it cannot be denied that he stinks … I do not blame the working man because he stinks, but stink he does.'

W. Somerset Maugham – *On a Chinese Screen*

One Monday morning, I visited Harrods, the enormous luxury store in central London. I had been told that early Monday morning was a good time to visit, as the regular customers would not yet be up and shopping, and the shop assistants would have time to talk to me. I started in the Food Hall, the basement in which expensive foods and drinks are sold, and I was given an extensive tour of the varieties of teas and coffees, each variety presented in a bucket-sized tin with an oriental look. At the tea counter I was invited to sniff Vintage Pu-erh, which had a tang of seaweed, and Hajua, which smelled more earthy; Mokalbari, which smelled earthy and honeyed with a hint of spice; and Aloobari, which smelled of heather on a Scottish hillside. I sniffed first flush Darjeeling, picked from young buds only and smelling brightly of capers and olives, and Darjeeling Castleton with a faint scent of muscatel grape, and Lapsang Souchong, a smoked tea, reminiscent of leather or old wooden ship.

Moving from the tea counter to the coffee counter, I was taken through the aromas of Mocha (chocolatey, smokey and bitter), Monsoon Malabar (smelling smokey and spicy), Hawaiian Kona (with a fragrance of caramel with citrus and thyme), Blue Mountain (nutty, caramel) and finally Kopi Luwak. Kopi Luwak is the most expensive coffee in the world. It is, according to the sales pitch, collected in the Garo Highlands of Sumatra where a species of civet cat eats the coffee cherry, digests the flesh, and excretes the bean. The bean has therefore been part-fermented in the guts of the civet. It sells at £2,000 the kilo. I was told by a coffee expert (not at Harrods, where I am sure they only sell the genuine article) that a great deal more Kopi Luwak is sold than is produced. 'Do you sell much?' I asked the Harrods sales assistant. 'Oh yes, the rich love things that are ridiculously expensive,' was the reply. 'The stinking rich', I commented. The sales assistant grinned: 'The wives of the stinking rich, with shitloads of money.' He added, 'You should go upstairs and sniff the most expensive perfume in the world.'

I took the escalator to the sixth floor of Harrods to sniff the most expensive perfume in the world. This floor had been modelled on the side chapels of a medieval cathedral. Each chapel housed the holy relics of a different perfume house, and each one had its sales priest to minister to the customer. There were good-looking sales girls, with tight skirts and elegant stockings; the sales boys were slightly more showy, and paid more attention to each other than to the customer.

I was third in line for the most expensive perfume in the world. Two Chinese boys, barely twenty years old, were ahead of me. They wore shorts and flip-flops. They had nearly finished choosing their personal fragrances, and it was time to pay up. They had chosen a relatively inexpensive perfume – a mere £405 per fluid ounce for each boy – though no doubt they would tell their friends a different story. They each pulled a wad of large-denomination notes from the back pocket of their shorts and peeled off a batch of the necessary.

Then it was my turn. I was introduced first to the economy version of the most expensive perfume in the world – £2,700 per fluid ounce. A bottle fitted comfortably in the palm of my hand. I was given a squirt of the scent on a card to sniff. It was definitely not a cheap smell. In fact, I found myself admiring it: it was restrained and rich, and I was to find, since I kept the card in my pocket for some time after, that it lingered pleasantly for days. The top notes were bergamot (a variety of orange whose blossom makes an essential oil), lime, thyme, cardamom, nutmeg, artemisia and pepper. The middle notes were heliotrope, jasmine, rose and Florentine orris – orris being the dried root of iris which acts as a fixative, with an earthy, leathery smell. The base notes were musk, ambergris, Tahitian vanilla, cedarwood and 50-year-old Indian sandalwood. 'Any synthetic ingredients', I asked? 'All perfumes have some synthetic ingredients, sir. Ours has less, but they don't tell me about them.'

I had however only genuflected at a less expensive version of the most expensive perfume in the world. Set high up on the wall of the pseudo-chapel, like a tabernacle in a real chapel, was a tiny bottle laced with gold and diamonds. One fluid ounce; price £143,000. No, they had not yet sold a bottle at that price, but once there had been a Middle Eastern princess who came close to buying. She had, however, wanted her family crest on the bottle, and the perfume house could not agree to that.

You could only admire the sales pitch. No one could ever compete with the most expensive perfume in the world. Should they try to compete, one would simply add another diamond to the bottle, raise the price, and stay ahead. Although no one might buy that bottle, plenty of the stinking rich, and plenty of the children of the stinking rich, and their wives and mistresses, would buy less-expensive versions, and would boast back home that it was the most expensive.

The perfumer responsible for this extreme luxury had taken over the former perfume house that had served Queen Victoria and her consort Prince Albert. In addition to selling the Most Expensive Perfume in the World, he had kept the formulae for the favourite perfumes of these restrained royal people and had recreated them. I was given a squirt of Queen Victoria's perfume '*Rose de Mai*'. I found it very safe, redolent of summer flowers with no trace of the animal or the sexy, as one might expect from this very correct monarch. Prince Albert's scent was recognisably a cologne – bracing, clean and faintly medicinal thanks to the presence of clary sage in the mix. Finally I was given a squirt of an altogether sexier perfume, in which the more aphrodisiac ingredients were Egyptian jasmine and tuberose, with peach and rhubarb somewhere in the mix.

Armed with scented samples of the most expensive perfume in the world, I was ready to experience another of the many meanings of smell. In many animals, smell is a form of social communication, one that identifies the members of a clan or pack, which distinguishes friend from foe, and which marks social status. So it is with humans too: we are offended by the alien smell of other tribes and other classes, and we are reassured by smells that mark people with the same social status as ourselves. In the big city, where we are shoulder to shoulder with thousands of unknown people from many different countries and backgrounds, is our sense of smell active in distinguishing one from another and putting them in their social or racial pigeonhole? And would my own smell of expensive perfume mark my social superiority?

My destination was at the opposite end of the social ladder to Harrods – one of London's covered markets. My journey to Tooting Broadway Market was by public transport, and I started on a bus. The bus was well-filled in the fag-end of the morning rush hour, with African office cleaners returning home from the early shift, Polish builders, Indonesian nannies, mothers with children, relaxed young people with unstressed jobs, students and heaven knows who else.

The bus smelt only slightly of diesel exhaust and disinfectant, and much more of people. But not a disturbing or unpleasant smell of people. A mild, humane smell, clean, not oppressive with artificial fragrances or sweat. More a breakfast smell, evoking bread or porridge, bodies not long from the warmth of bed, domestic cosiness. Nothing unpleasant; nothing that smelt alien; nothing that evoked differences of class or race. A few people gave off a more noticeable smell – a man with the acidic, tomato smell of sweat, a waft of sour nicotine from someone who had recently stubbed out a cigarette, a young man with damp hair and extravagant stubble still smelling of shower gel. No doubt the experience would have been different in the rush hour crowds. Certainly a century ago, or less, it would have been very different. That was the era when the rich had acquired habits of cleanliness and the working man had not.

Somerset Maugham spent the winter months of 1919 travelling in China, and his travel book about the journey contains a long reflection on class consciousness and the smell of people. It is a masterpiece of snobbism, and it is free of irony; he really means it. 'I asked myself,' he wrote at the end of a day of travel in central China,

> why in the despotic East there should be between men an equality so much greater than in the free and democratic West, and was forced to the conclusion that the explanation must be found in the cess-pool. For in the West we are divided from our fellows by our sense of smell … I do not blame the working man because he stinks, but stink he does. It makes social intercourse difficult to persons of sensitive nostril. The matutinal tub divides the classes more effectually than birth, wealth or education … Now the Chinese live all their lives in the proximity of very nasty smells. They do not notice them. Their nostrils are blunted to the odours that assail the Europeans and so they can move on an equal footing with the tiller of the soil, the coolie, and the artisan. I venture to think that the cess-pool is more necessary to democracy than parliamentary institutions. The invention of the "sanitary convenience" has destroyed the sense of equality in men. It is responsible for class hatred much more than the monopoly of capital in the hands of the few.'

The fact that Somerset Maugham could write so disparagingly about the stinks of working men, and that he could be published, suggests that his point of view was not an isolated attack of snobbism, but was also widely appreciated amongst people who bought his books.

There is a delightful putdown from Mark Twain to the notion that prevailed at the start of the 20th and end of the 19th centuries that the working classes stink. Twain wrote an open letter in reply to one that had been published in a New York newspaper in 1870 by a Reverend T. De Witt Talmage of Brooklyn. The Rev De Witt Talmage's letter had contained the following utterance on the subject of smells: 'I have a good Christian friend who, if he sat in the front pew in church, and a working man should enter the door at the other end, would smell him instantly … If you are going to kill the church thus with bad smells, I will have nothing to do with this work of evangelization.'

Mark Twain published a long and damning reply:

> We have reason to believe that there will be labouring men in heaven; and also a number of negroes, and Esquimaux, and Terra del Fuegans, and Arabs, a few Indians, and possibly even some Spaniards and Portuguese. All things are possible with God. We shall have all these sorts of people in heaven; but, alas! in getting them we shall lose the society of Dr. Talmage. Which is to say, we shall lose the company of one who could give more real 'tone' to celestial society than any other contribution Brooklyn could furnish. And what would eternal happiness be without the Doctor? … If the subject of these remarks had been chosen among the original Twelve Apostles, he would not have associated with the rest, because he could not have stood the fishy smell of some of his comrades who came from around the Sea of Galilee. He would have resigned his commission with some such remark as …: 'Master, if thou art going to kill the church thus with

bad smells, I will have nothing to do with this work of evangelization.'

Well into the 20th century the smell of the lower orders was still nagging at the minds of the middle classes. George Orwell was thoroughly on the side of the poor, and spent some months in 1937 in the slums of Lancashire, experiencing their down-trodden lives. Despite his sensitive social conscience, he was deeply upset by the smell of the 'lower classes'. For him, the bad smell of the lower classes mattered more than racial, religious or educational differences.

We should count ourselves lucky that the world has moved on. Class divides certainly continue, but they are no longer reinforced by the idea that 'the lower classes smell'. Although it is difficult to be sure about class notions in the distant past, it seems likely that smells were a factor in the class war only for a period from the late 18th through to the mid 20th centuries, when the moneyed classes washed and scented themselves, while the working classes sweated. Before that, in a largely agricultural society, it seems likely that almost everyone shared agricultural smells of sweat and animal. Deodorants, toothpaste and washing machines have done for the stink of the lower orders, though at a price. The price we pay for hygiene and deodorants is in the pollution pumped out by a billion washing machines and the pollution of soap, toothpaste and washing powder flowing down to the seas. And perhaps there is a new way in which smells reinforce the divisions between people, as the smell of deodorants and washing powders that seem inferior activate our defence mechanisms against the alien.

My bus dropped me at an Underground station. Through the concourse, past the tangle of tourists and wheelie suitcases that had become jammed at an unfamiliar ticket barrier, down into the real heart of the Underground where the air was rich with the ancient smell of tube trains. This was the smell that I had really hoped to find, and it needed careful analysis. It is a smell unique to underground railways, to be found nowhere above ground. No doubt it is similar in the metros of other cities, though many include rubber in the blend of odours, whereas London's tube trains have metal wheels. It is a hot, dusty, electric, oily, mealy, sooty, earthy, metallic smell with many layers in the blend. There is a dominating note of warm lubricating oil, mixed with dirt: a smell that reminded me of old-fashioned metal-bashing factories. There is a fruity note akin to fig with a bitter edge that probably derived from brake fluid. There was a smell of baking that I could only suppose was the combination of heat with dust and with underground fungi, probably a product of the Maillard reaction, and maybe of fermentation.

Changing from the Underground at Victoria Station, I joined the tide of people flowing towards the overground trains. Victoria Station was clean and airy, and given over to a growing number of neat little shops, each with its own smell. Coffee, croissants, cinnamon, butter, the sharp clean smell of magazines like fresh paint. Some of these smells were suspiciously artificial, and were being pumped out into the concourse to encourage passers-by to pause and buy. This was ambient scenting, which we have already encountered: the use of synthetic aromas to draw in customers and to create an atmosphere conducive to spending.

Then there was an authentic, overpowering English aroma. It was the Cornish Pasty Company, with a smell that is almost a public nuisance, staining the air for yards around with the sourness of onions, overlaying and underpinning the blunt smell of meat with bland potato and lardy pastry. The Cornish Pasty Company in Victoria Station was punching above its weight that morning: one Latin American woman was managing the counter but her stall was contributing the strongest smell of the station. It was sucked along with the flood of people to the platforms.

A railway carriage of an overground commuter line must carry several thousand packed people every working day, and the smell of this carriage was heavy-duty. It smelled of sour, unclean synthetic carpet and unclean synthetic plastic flooring, and sour, synthetic-flavoured disinfectant. It was a grey smell – not quite dirty, but not clean; not quite unnatural but not related to anything that grows or breathes; not quite industrial but not remotely domestic. In the mix there was a good dose of lubricating oil, a medium dose of sad newspaper smell, and a small dose of the bready smell of morning people.

My destination was Tooting Broadway covered market, at the opposite end of the social range to Harrods. The covered markets of London are an olfactory joy. There are high-quality markets like Borough Market where the dominant smells are safe and luxurious – lilies, olive oil, Parmesan, salami, strawberries, truffle oil. Then there are other markets, like my local Tooting Market, where the smells are less guaranteeably safe, more foreign.

We are primed to be disgusted by foreign smells. Foreign smells often mean dodgy smells, redolent of unhygienic practices. We approach them nervously, unsure whether to breathe freely in case we encounter something unwholesome. Worse, there is a tendency for our disgust at foreign smells to seep out beyond the smell to the people: so intolerance of smells becomes intolerance of foreigners, and no doubt contributes to the occasional bouts of isolationism and xenophobia into which the human race retreats. Stepping nervously into Tooting Market, I was in for a surprise. There were foreign smells aplenty, but not unpleasant, and if anything less threatening than the high-pitched blast of perfume in Harrods.

There was a Chinese halal kitchen at the entrance to the market, giving off the clinging smokiness of cooking oil, blended with the vinegar and toffee scent of sweet-and-sour sauce. There were sari stalls, smelling of old-fashioned camphor mothballs. There were stalls selling joss sticks smelling of frankincense, ylang-ylang and vanilla. A stall selling African materials in saturated West African colours gave off a dusty rough smell of cotton. A Portuguese grill smelled of hot cooking oil, leather, fried fish and vinegar. A pet shop gave off a seedy, old straw smell mixed with sour bird shit and leather. On closer investigation, I found the surprising smells of dog chews made of tripe, beef hide and pig's ear, each with its distinctive aroma. A Mauritian foodstall offering the food of four continents – pattie, *feuilleté*, *roti* and biryani. A Chinese herbalist let me sniff his many jars of dry things and savour the variety of herbs – cool or dusty, mealy or honeyed, citrus or tobacco, hay or nut.

It was not these foreign smells that gave me pause, but the familiar, puzzling smell of a butcher's stall. I have never understood why a butcher's shop should smell not at all of live animal nor of cooked meat, but a sour, fresh smell not too distant from the smell of a fishmonger.

The butcher I encountered showed himself to be an articulate observer of smells. 'Pork is vinegary and sort of pissy. Smell my fingers' (he was right – a distinctly sour smell that bears no relation to cooked pork or bacon). 'Beef can get pretty pissy. It depends on the day – a day like today, muggy, everything is worse. The smell lingers on you; you take it home; horrible.' And lamb? 'Smells like a baby lamb, sort of woolly and sweet.' And goat? 'Sweet and pissy.' And the blood, isn't that a distinctive part of the smell? 'No.' And the sawdust? 'Only when it gets damp.' So that was it – at last I had a satisfactory account of the smell of a butcher's, a smell that had puzzled me for much of my life. The smell of butchers' shops is not the blood, not the sawdust, not some secret chemical in the chillroom. It is the pissy, vinegar smell of raw pork overpowering everything else and asserting itself over the smells of neighbouring stalls.

A friendly Jamaican man was running a lunch stall nearby, and he called me over. 'What you taking notes for?' I explained my interest in smells and he began testing my powers, offering an empty frying pan for me to sniff and say what I could detect. I began cautiously – 'I can smell onion (good), vegetable oil (OK), batter (very good), and a herb, or perhaps a fresh-smelling spice, maybe turmeric?' ('Yes man, I put turmeric in everything'). I had passed the test. The next pan he passed for me to sniff was not a test but a shared exploration. I could smell and identify the onions, coconut, rice, fish and beans. However, I could not name the capsicum peppers, nor the ackee, which were also in the recipe, nor the Scotch bonnet pepper which he would have included if he had had a Scotch bonnet pepper. But never mind, we were friends.

My journey across the social divides of London had taken me from the fragrance of wealth in Harrods to the aroma of turmeric and coconut at a Jamaican fast-food stall. It had not, however, shown that wealth smells better, nor that 'foreign' smells worse. There was pleasure to be had at both ends of the social scale. I had experienced plenty of varied smells, but I had not been offended or attracted by smells that had powerful social meanings. Instead I had simply enjoyed the variety of life in a big city. Knowing that my ordinary nose is capable of sniffing its way through such variety, and that my ordinary brain is capable of learning to make use of this elusive sense added to the pleasure. Learning is in large part a matter of creating hooks in our minds on which we can hang new knowledge, and we can create these hooks for new knowledge about smells. Words help, and the next chapter will explore how writers have used words to convey smells.

CHAPTER 15

Holy Smells, Pious Smells

'"Sancho, it strikes me thou art in great fear."

"I am," answered Sancho; "but how does your worship perceive it now more than ever?"

"Because just now thou smellest stronger than ever, and not of ambergris," answered Don Quixote.'

Miguel de Cervantes – *Don Quixote*

Many writers sidestep the sense of smell as being too difficult to describe, or too vulgar. For a short period in the history of literature, however, the sense of smell was a permissible subject for serious writing. This period covered the second half of the 19th century and the very first decades of the 20th, when realism and symbolism were in vogue. Realist writers wanted to capture the full sensual experience of lives and places – often of particularly gritty lives and places – and found that smells were an essential part of that experience. Symbolist writers often wanted to convey the secret connections between things, and found that odours offered a way into the hidden structures of life.

I was brought face to face with a symbolist view of smells when I visited a garden in France, and found a poem tacked up at the exit. It was a garden devoted to perfumed plants and it offered a more staggering range of natural scents than

I have ever encountered. The poem was by Baudelaire, the great French symbolist poet, and was published in 1857. It is called 'Correspondences', and it is about the sensory language of nature, and how scents, colours and sounds talk together in harmony:

> Nature is a temple where living pillars
> Sometimes give voice to confused words;
> Man crosses it through forests of symbols
> Which watch him with intimate looks.
>
> Like those deep echoes mingling in the distance
> In a dark and profound harmony,
> Vast as the night and as the clear of day,
> So perfumes, colours, tones answer each other.
>
> There are perfumes fresh as children's flesh,
> Soft as oboes, green as meadows,
> And others, corrupt, rich, triumphant,
>
> Within the vast range of infinite things,
> Like amber, musk, incense and aromatic resin,
> Chanting the ecstasies of souls and senses.
>
> Charles Baudelaire – *Les Fleurs du Mal*

I had never expected much from a sensory garden. Such gardens are meant to appeal to all the senses – sound, smell and touch as well as sight – but they frequently disappoint. I had visited several which rest their claims on a few scented roses, some lavender bushes, and bamboo chime bars. But at the *'jardin aux plantes parfumees'* of La Bouichere at Limoux in the foothills of the Eastern Pyrenees, I was amazed. It is not a grand garden; it is to be found amongst the tatty warehouses and superstores with which the French surround their towns. Past a supermarket, a mattress warehouse, a DIY store, and there is the modest entrance, marked with a line of pots of pelargoniums, which belong to the geranium family. I bent to a pot, rubbed a leaf between finger and thumb and sniffed.

An hour later I had barely finished with the pelargoniums – 60 plus different varieties, and never a repeated scent. Two hours later, I had just about navigated the family of mints and of basils and was deep into the families of plants whose scents were complete novelties to me. Each family was like a fuse of varied scents that led from one spark to the next; no family of plants repeated itself. Flowers may loom large in the world of perfume, but a wander around La Bouichere showed me that the scent of flowers is knocked into a cocked hat by the variety of leaf smells.

To start with the pots of pelargoniums at the entrance to the garden. These are the scented geraniums, first brought to Europe from South Africa. *Pel Lady Scarborough* smelled of a blaring orange fruit gum mixed with a subtle strawberry. Its neighbour, *Pel Sweet Mimosa* had a flower that smelled of rose, and a leaf that was fresh cut grass. The next plant laboured under the name *Cola Bottles*, and its flower was indeed cola, but its leaf was the fragrant gentian scent that comes with the drink Suze or Angostura bitters. Next was a pelargonium that was all orange: a delicate scent of orange blossom from its flower and a gingery orange peel from its leaf. *Pel Bontrosai* was a mix of lemon and pine. *Pel Blandfordiana* was described on its label as camphor-sage, and was to my nose a full-throated blast of industrial cleaner. The leaves of *Pel Filicifolium* had a strong scent of ripening banana.

More scented geraniums. In the order in which I sampled the leaves, I encountered Turkish delight, plum, a hefty mix of peanut butter and maple syrup, cider apple, incense and citrus mixed, a chalky marine smell like the sea at Dover, old-fashioned hair cream (from a plant called 'Brilliantine'), citronella, absinthe, the eucalyptus smell of Vicks rub, a mix of Virginia tobacco and lemon, parsley, a peppermint much stronger than the peppermint scent of peppermint (more like the powerful cold remedy called Fisherman's Friend), black pepper, coconut, cider, cinnamon, the gentle wood smell that the French call '*boisé*', mint chocolate, faded carrot, a mix of mandarin and ginger, lavender, resin, hyacinth, green apple, apricot, a mix of pine and fruit gum, cedar, fresh cut grass.

This staggering variety is all to be found in the leaves of pelargoniums, which you can savour by gently squeezing the underside of the leaf between thumb and first finger. The resulting scent is not only crazily varied from one plant to the next, it is also a multi-layered scent that evolves as you sniff it. The first whiff is usually assertive and volatile, and it attacks the nose in the first split second of sniffing. Once those aggressive molecules have evaporated, the scent evolves through several gentler changes. This is the effect that perfumers seek when they construct the head notes, the middle and bass notes of a perfume. But nature got there first, and at a fraction of the cost.

One scented geranium is special to perfumers. *Pel graveolens* yields the essential oil called 'attar of roses', a perfume that used to stand for vulgarity in the perfume world. Shady foreigners in early spy novels might scent their handkerchiefs with attar of roses. Amongst the long line of fragrant plants at La Bouichere, though, it hardly stands out.

A shocking plant introduced me to a family that I had never considered as a source of smell. It was *Coleus Canina*, popular name 'scatter-cat'; the Latin name is perhaps a naturalist's joke, as dogs are the sworn enemies of cats. It was a knock-out whiff of concentrated lemony garlic. Next to it was *Plectranthus amboinicus* – from the same botanical family – smelling cool and herbal, something like oregano. And *Plectranthus tomentosa* ('Vicks plant'), smelling like an old-fashioned lavatory cleaner. The family that includes coleus and plectranthus is another family of plants like pelargoniums whose chemistry produces a variety of powerful smells that ranges from the sweetest to the foulest, from the fruity to the meaty, from floral to lavatorial. Then I came across a complete novelty – *Mertensia maritima* – which smelled of oysters and mushrooms. Here was a flower that smelled of milk chocolate – *Berlandiera lyrata*. Here was a *Melianthus* whose leaves smelled of Marmite (though some say peanut butter). Here were ginger lilies whose flower is conventionally fragrant, and whose root gives us ginger. Then there were the mints. Each had a cool minty note coupled with other wildly varied layers of scent. The variety included goat, sponge cake, tomcat, banana, eau-de-cologne, grapefruit, pineapple, eucalyptus, apple. I noted the subtle differences between the Italian mint, the Algerian, the Moroccan and the Corsican mints.

Why such variety? Aromatic plants are of course found not only around the Mediterranean Sea, but in all the Mediterranean climate regions of the world. As well as the aromatic plants of hillsides around Mediterranean Sea – the lavenders, thymes, cistus – there are the eucalyptuses of Australia, the scented pelargoniums of South Africa, the Chilean myrtles such as

Luma chequen whose leaves are raspberry scented, and the many sages of California. The production of essential oils is an interesting strategy by which aromatic plants adapt to the difficult conditions of the Mediterranean climate. The primary role of essential oils is to protect the plant against predators – grazing animals or insects. A second, more complex, role of essential oils is to combat competition from other species: the litter of decomposing leaves formed beneath cistus or thyme plants, for instance, releases substances that inhibit the germination of competing plants.

Although scientific studies are still underway, it seems that there is a third reason why these plants are aromatic: air conditioning. It would appear that the essential oils in their leaves may also protect the plant against dry conditions and strong solar radiation. Essential oils are produced by special organs, the glandular hairs, on leaves. Even more effectively than the simple hairs on grey-leaved plants, the dense web formed by these glandular hairs traps a small film of air on the leaf surface which serves as a barrier against the air outside, and in so doing, limits water loss from the leaf's pores. The evaporation of essential oils also has a cooling effect on the environment immediately around the plant, protecting its leaves from burning. In periods of extreme heat, therefore, the evaporation of essential oils may be very significant, hence the unique scent of Mediterranean vegetation in summer. In California, in the hills around Hollywood, the sage *Salvia apiana* gives off such a quantity of essential oils that it is visible as a mist floating above the chaparral. The stronger the plant's scent, the more it is protecting itself against heat and sun. Their essential oils enable aromatic plants to shelter within an air-conditioned bubble – a private atmosphere created around each plant.

The garden of La Bouichere displays nature as a wildly inventive perfumer might: creating an endless series of variations on the theme of geranium or mint. It is fitting that a poet like Baudelaire, who saw the hidden meanings of nature's works, should be chosen to celebrate such a profusion of perfumed plants. It would be hard to find an English garden using an English poet to celebrate this aspect of nature. The English are not so much at ease with the sensual as are the French. When it comes to writers, there are many French authors who deal well with smells, and very few English ones. There is even an official count of smell references in French literature:[28] from which the novelist Émile Zola comes out top.

To take one example, Zola's novel *Le Ventre de Paris* is located in Les Halles, the great wholesale market of Paris. The book dwells lovingly on the smells of produce and people. Here for instance is Zola on the cheeses to be found on sale in Les Halles:

> The *camembert*, with its gamey stink ... smothered most of the other smells under a waft of bad breath. However, in the middle of this powerful phrase, the *parmesan* threw in a thin, flute-like note, while the *brie* played with a faded sweetness like faint tambourines, the *livarot* gave a short, suffocating reprise and this symphony paused for a moment on the sharp note of fennel-scented *géromé*.

Here again we find the connection between music and smells that perfumers find so natural – smells seem to communicate to us in notes, keys and chords.

But for delicacy and accuracy of observation, there is no

French writer to beat Proust. We have already encountered him writing about a madeleine dipped in herb tea in the rooms of Aunt Léonie, and the memories sparked off by the cake's flavour. A few pages after that famous 'madeleine' moment comes a passage conveying the smell of Aunt Léonie's rooms. I defy anyone to write a better, more vivid and detailed description of indoor aromas than this:

> My aunt now lived effectively in only two adjoining rooms, passing the afternoon in one while they aired the other. These were the kind of provincial rooms which (just as in certain countries whole tracts of air or sea are illuminated or scented by a myriad of phosphorescent organisms which we cannot see) delight us with the thousands of odours breathing out into the atmosphere their virtues, their wisdom, their habits, a whole secret life, invisible, overflowing and moral, to be held in suspense in that atmosphere; natural smells certainly, and coloured by the weather like the countryside outside, but already humanised, homely, confined, an exquisite jelly, well-made and limpid, blending all the fruits of the year which have left the orchard for the store-room; changing with the seasons, but established and domesticated; balancing the sharpness of hoarfrost with the sweetness of warm bread, leisurely and punctual like a village clock, sauntering and settled, thoughtless and provident; linen smells, morning smells, pious smells; rejoicing in a peace which brings only an increase in anxiety, and happy in a prosiness which serves as a deep reservoir of poetry for those who pass through it without being part of it.

We can picture Proust as a small boy, savouring the bourgeois smells of his family home and framing this description in his mind to be revived later in adulthood. More likely, we should picture Proust the writer, lying in bed in his Paris flat (afflicted by asthma, and bed-ridden like Aunt Léonie), bringing back to the forefront of his mind the olfactory memory of his aunt's rooms. There he lay, his bedroom lined with cork to deaden the sounds of Paris, searching his mind for time past. Did he have the description ready-framed in his head? Or was the memory sparked off by some smell in his immediate surroundings? Or did he imagine what the memory would be like, if he had been able to resurrect it at will? Most of us cannot summon up an olfactory memory at will, and Proust himself described his smell memories as 'involuntary'.

Looking beyond France for writers who can deal with smells as well as Proust is a difficult task, but not a hopeless one. A great majority of writers describe a world of genteel or aristocratic life in which smells have no place. A very few writers have opted for an extravagant, Gothick approach to smells. Amongst these are Virginia Woolf (whose book *Flush*, which I have already mentioned, features a dog), and the German writer Patrick Süskind whose many million-selling novel *Perfume* features a psychopathic murderer with hyperosmic powers. For these writers, their imagined world is redolent with smells that are thick, clotted and coloured, and that bear no resemblance to the normal experience of olfaction.

Then there are writers who throw in a few smells in a nonchalant way to provide some evocative detail. There are many such writers – Ernest Hemingway is a good example: his novels contain some evocative smells, which serve well to bring a location to life, but which are never described for real. So, in

A Farewell to Arms, the protagonist, on the run from war, arrives in the Italian city of Milan early one morning: 'It smelled of early morning, of swept dust, spoons in coffee-glasses and the wet circles left by wine-glasses.' Or, 'there was the smell of marble floors and hospital'. Or 'the hay smelled good'. These seem on the surface vivid details, but they do not really describe smells, they do not convey the smell of early morning or of marble floors or of hay.

For a writer in English who can describe smells as well as Proust, and who understands their power to convey emotional states, one has to turn to James Joyce. Joyce is the only writer who has captured the very act of smelling: its uncertainties, its repeated sniffs, its hesitant attempts to pin down a smell by association with other remembered smells. In the vast novel *Ulysses*, the protagonist Leopold Bloom at one point checks his own body odour: 'Mr Bloom inserted his nose. Hm. Into the. Hm. Opening of his waistcoat. Almonds or. No. Lemons it is. Ah, no, that's the soap.' (This is the soap – 'sweet lemony wax' – he bought for Molly his wife 300 pages before, and which accompanies us to the end of the book.)

I stand in awe of the words Joyce uses for smells. 'The sweet oaten reek of horsepiss'. Babies: 'sour milk in their swaddles and tainted curds'. Mansmell: 'celery sauce'.

I stand in further awe of Joyce's ability to make smells the means to evoke the emotional states of his characters and to involve the reader. Look at the very first page of *A Portrait of the Artist as a Young Man* – 'When you wet the bed first it is warm then it gets cold. His mother put on the oilsheet. That had the queer smell. His mother had a nicer smell than his father.' It is plain that Joyce wants us to experience being small again through the sense of smell. The small boy goes to school, and the smell of school chapel is his first memory of it. 'There was a cold night smell in the chapel. But it was a holy smell. It was not like the smell of old peasants who knelt at the back of the chapel at Sunday mass. That was a smell of air and rain and turf and corduroy.' Then comes the young artist's struggle with religion, tainted with smells of increasing doubtfulness: 'to drink the altar wine out of the press and be found out by the smell was a sin too: but it ... only made you feel a little sickish on account of the smell of the wine.' And the smell of religion reaches its high point in a religious retreat, where a ranting Jesuit depicts the horrors of hell:

> Consider then what must be the foulness of the air of hell. Imagine some foul and putrid corpse that has lain rotting and decomposing in the grave, a jellylike mass of liquid corruption. Imagine such a corpse a prey to flames, devoured by the fire of burning brimstone and giving off dense choking fumes of nauseous loathsome decomposition. And then imagine this sickening stench, multiplied a millionfold and a millionfold again from the millions upon millions of fetid carcasses massed together in the reeking darkness, a huge and rotting human fungus. Imagine all this and you will have some idea of the horror of the stench of hell.

'A moral odour', 'a holy smell', writes Joyce; 'pious smells ... breathing out their virtues ... their wisdom', writes Marcel Proust. That smells should convey morality, wisdom and holiness is a new and difficult concept. I decided I should explore the smells of holiness in a country church. A visit to an old country church in England is an experience in a minor key. The 'chord' of smells is muted, the notes are low and earthy, the underlying sense is solemn and slow, more ancient than one can quite grasp. The smell is old, earthy and rural, with none of those ecclesiastical touches that come from incense and candles and perfumed people, and might be found in an Italian or Spanish church. It is holy in a homely way, not an exalted way. It leans towards the ground, not towards heaven.

Just outside the porch, the church smell reached me. It was less a smell with identifiable layers than a sensation, a sensation of past feet on stone, past hands on wood, past lives and deaths. It seemed very like the sensation of swimming out from a beach and reaching a margin of colder water that warns one of great depths and distances.

It is a smell, not exactly of stone, but of an interior that has long held dust from stone and earth. It is musty, not the mustiness of closed-up rooms or of uncared-for spaces, but old surfaces that have never fully warmed up. To the old smells of earth and cold stone, there is added the leathery aroma of pews and hymnals that have been handled by generation after generation of church-goers, the faded vegetable smells of tired arrangements of gladioli, and the faint woody smells of church leaflets and Sunday school cardboard.

Afterwards, as I stepped out of the church into fresh air, I wondered about the words to describe the smell of fresh air. Later, I came across a passage written by a German poet that seemed to capture it perfectly. The poet was Rainer Maria Rilke, and his book *Letters on Cézanne* is a compilation of letters he wrote from Paris to his wife in North Germany. They had been corresponding almost every day throughout September 1907, and she had sent with her last letter three small sprigs of heather from her local North German countryside. He wrote back:

Paris IVe, 29, Rue Cassette,
September 13, 1907 (Friday)

… Never have I been so touched and almost gripped by the sight of heather as the other day, when I found these three branches in your dear letter. Since then they are lying in my Book of Images, penetrating it with their strong and serious smell, which is really the fragrance of autumn earth. But how glorious it is, this fragrance. At no other time, it seems to me, does the earth let itself be inhaled in one smell, the ripe earth; in a smell that is in no way inferior to the smell of the sea, bitter where it borders on taste, and more than honeysweet where you feel it is close to touching the first sounds. Containing depth within itself, darkness, something of the grave almost, and yet again wind; tar and turpentine and Ceylon tea. Serious and lowly like the smell of a begging monk and yet again hearty and resinous like precious incense. And the way they look: like embroidery, splendid; like three cypresses woven into a Persian rug with violet silk (a violet of such vehement moistness, as if it were the complementary colour of the sun). You should see this. I don't believe these little twigs could have been as beautiful when you sent them: otherwise you would have expressed some astonishment about them. Right now one of them happens to be lying on dark-blue velvet in an old pen-and-pencil box. It's like a firework: well, no, it's really like a Persian rug. Are all these millions of little branches really so

> wonderfully wrought? Just look at the radiance of this green which contains a little gold, and the sandalwood warmth of the brown in the little stems, and that fissure with its new, fresh, inner barely-green. – Ah, I've been admiring the splendour of these little fragments for days and am truly ashamed that I was not happy when I was permitted to walk about in a superabundance of these. One lives so badly, because one always comes into the present unfinished, unable, distracted.

Rilke's idea that, by being attentive to small smells and the look of small things, we can be happier and live better in the present, seems to me to be a fine by-product of this exploration of the sense of smell. By paying close attention to smell, we are forced to interrupt the usual calculations and conversations in our heads, and to live for a brief moment undistracted in the present.

CHAPTER 16

The Smell of Three O'Clock in the Morning

My son Nick has an unusual mind, and an unusual question about smells. He was born blind, and in the 32 years since his birth has accumulated many labels of disability – autism, learning difficulties, epilepsy, behaviour problems. But all this pales into insignificance in the light of his outstanding talent for music. Mozart started to play the piano at the age of three; Nick beat Mozart by two years. He taught himself to play when he was still in nappies, and had to raise his hands well over his head to reach the keyboard. From that early age, he has shown a serious, intellectual grasp of serious, intellectual music. He cannot add two and two, but he loves composers like Bach, Chopin, Debussy and Shostakovich, and he plays their music with deep understanding.

His unusual question about smells is 'What does three o'clock in the morning smell of?' It is a puzzle to know what is behind this query. Nick finds language fairly difficult, and struggles to put his inner sensations into words. Most of us are asleep at three o'clock in the morning, and the sense of smell goes to sleep with us. Most of us dream without images of smells; though there have been studies that show that roughly one in four of us have occasional dreams in which a smell or, (much the same thing), a taste feature. Nor do real smells impinge on our sleep, though there is research that suggests that smells can influence our dreams. Researchers at the University of Dresden have, for instance, experimented with exposing volunteers to pleasant and unpleasant smells during sleep, and investigating the dreams they report under the influence of such odours. The good smell they chose was phenethyl alcohol, which smells of roses; the bad smell was hydrogen sulphide which stinks of rotten eggs. Participants reported more positive dreams with the good smell, and more negative dreams with the foul smell, but they did not report that smells featured in their dreams.

So it may be that Nick dreams at three o'clock, and that his dreams have an olfactory tinge. Helen Keller, who was deaf and blind, reported that in sleep 'I smell and taste much as in my waking hours. In my dreams I have sensations, odours, tastes and ideas which I do not remember to have had in reality.'

Another possibility is that Nick has some form of synaesthesia. His head is full of music most of the time, and the music may connect for him with some olfactory sensations. He says, for instance, that 'Bach smells sort of like a rat. An oboe smells of bleach. *Romeo and Juliet* (the ballet with a score by Prokofiev) smells of wine, by the way. *The Nutcracker Suite* (by Tchaikovsky, a favourite of Nick's) smells sort of wonderful, sugary.' Yet another possibility is that epilepsy has something to do with the smell of three o'clock. For some people with epilepsy, seizures announce themselves with phantom smells, often the particularly acrid smell of burning toast. Perhaps, for Nick, an epileptic seizure while he is asleep has announced itself with such a phantom smell.

At a loss to know how to answer Nick's question, I decided to take him for an exploration at three o'clock in the morning to some of the places in London that are active and smelly at that early hour – the great wholesale markets. We woke, breakfasted lightly, and set out into the dark. Outside the front door, we paused and sniffed. 'What does it smell of?' I asked. Nick often answers questions with questions: 'Does it smell of the gym, sort of sweaty? Does it smell of foxes?' he asked. I could smell neither gyms nor foxes. The air was cold, and my nose was chilled. There was a faint hint of rubbish bin in the air, a touch of trodden autumn leaves, both perhaps a little foxy or sweaty. But to me, the prevailing sensation was of the strangeness of the hour, and the absence of the usual sensations of a big city. It was a strangeness compounded of the deadening silence, the cold, disorienting fog making the streets look unreal and theatrical, and the absence of the smells of activity – no car exhaust, no people smells, no cooking smells. A city of millions of people seemed to be inhabited only by traffic lights going through their robotic routines with no people to control.

Our first port of call was Billingsgate, the great fish market of London. We drove through the rain-slicked streets, seeing only an occasional delivery lorry and an occasional fox. Urban foxes, trotting with stiff legs and stiff tails, always at right angles to human traffic down alleyways, through hedges, under crash barriers, became more numerous near the big markets. 'Do you know what a fox is like?' I asked Nick. 'Foxes are big, aren't they? Like a horse, by the way. They come indoors. They pant.' I put him right about the size of foxes, and their potential threat to people. And what of the smell of foxes? 'Foxes smell tense,' said Nick, 'sort of like Shostakovich, by the way.' His comment seemed to me surprisingly perceptive, as foxes have a very edgy stink, and most of the music composed by Shostakovich has a tense feel. 'A fox ate Christmas Pudding,' said Nick. He was recalling a moment from childhood. The school he attended had a small menagerie of animals, including a cantankerous rabbit called Christmas Pudding. For the holidays, the animals were farmed out to willing pupils. Nick got the rabbit, and we put its hutch, whose latch seemed secure, in the garden. One night a fox picked the lock on the rabbit's hutch, and left the gruesome remains of Christmas Pudding scattered around the garden. Perhaps it was this memory that accounted for Nick's question about three o'clock in the morning.

The streets approaching Billingsgate fish market were empty and seemed particularly inhuman in the early morning – tunnels, underpasses, places with numbers and no names. But on arrival, we found a different world. The mighty warehouse which contains the market was full of people, bright with light, cheerfully noisy with shouting, and glittering with ice.

The market did not stink; it had a mild fishy smell, metallic and bitter, to which we rapidly became inured. I wanted to know about the varieties of fish smells, but with no success. A stall holder explained: 'It's the ice; chills your nose. You would notice the different smells if you were gutting the fish … they smell different when you gut them. But here they just smell of fish. Fishy.' Nick chipped in: 'Ducks smell of fish, isn't it? And churches: they smell of damp. Musty. Ammonia – that's a funny smell sometimes in the back of my skull, by the way.'

'It seems to me, young man,' said the fish wholesaler to Nick, 'that you are only interested in bad smells.' Nick replied: 'Smells are like all the memories, aren't they? Smells are like houses. Our house smells of mice. The house where I had my music lessons had a very polishy smell. Irene's house smelled of old lady. Grandma's house smelled of bread, by the way. The smell I didn't like was Myrtle's house: it smelled of cats. I like the fresh air smell: it's sort of lovely. It's like Debussy.' That concluded the conversation. None of us were up to discussing the smell of Debussy with Nick, and the fish man had clients to serve.

Our next stop was Smithfield, the meat market. A livestock market has been held on the present-day site, next to the financial heart of London, since the 12th century. In 1174 it was described by William Fitzstephen, clerk to Thomas à Becket, as a 'field where every Friday there is a celebrated rendezvous of fine horses to be sold, and in another corner are placed vendibles of the peasant, swine with their deep flanks and cows and oxen of immense bulk.' Bulk is the continuing theme of Smithfield. The lorries that deliver meat to the wholesalers are parked in line like oxen of immense bulk. The wholesalers themselves are serious and bulky. The sides of meat hanging up in refrigerated areas behind transparent plastic curtains are deep-flanked and bulky. As for smell, the market is a disappointment: all the meat is refrigerated and, even if it were not refrigerated, would hold little interest for the nose.

Raw meat, like raw fish, offers very little variety to the nose until it is cooked. One can distinguish fresh from tainted meat, but I for one am not attuned to the differences between raw pork, beef, lamb, venison or mutton until they are cooked. One might have expected the human nose to be good at discriminating between such important items in our diet, in the way that it is good at distinguishing the various aromas of berries and fungi and herbs and nuts. Perhaps the explanation may lie in our distant past, when our ancestors were hunter/gatherers.

The men did the hunting for meat, and relied on sight for spotting prey and for reading their tracks; the women did the gathering, and had to use their noses as well as their eyes to find food. That is a hypothesis to explain why women are typically more sensitive than men to smells, and better at identifying smells. Like so many of these speculations about our evolutionary past, it is unprovable, but the hypothesis could be placed alongside Professor Thomas Hummel's that women are better than men at smelling because they are more interested in the social messages that smells convey.

Our final stop was the fruit, vegetable and flower market at New Covent Garden, on the South Bank of the River Thames. It is an immense warehouse built on former wasteland between Victorian railway lines. On the outside there is nothing to hint at the profusion inside, and it takes some searching among vans, forklift trucks and waste bins to find a way in. But inside there is a series of covered alleys in which the wholesalers display their fruit and vegetables and do their business. What a display it is! Everything is colourful, everything is redolent, everything is displayed in regimental order in fruit boxes. There is every colour of tomato, every colour of carrot, every colour of brassica, and several colours you have not seen before. There are ranks of figs, cabbages, artichokes, oranges, fresh herbs, fennels, broccolis, hazelnuts, all displayed like freshly painted toy soldiers.

Walking down the alleys, our noses inspected the smells of each wholesaler's specialities as we passed. Nick gave a running commentary in his usual uninhibited way: 'Oranges, they smell Christmassy. Cabbage – that's a yukky smell, kind of cellarish. Bananas: they didn't have bananas in the World War (from his history lessons this one fact has stuck in Nick's brain). Pineapples, I know pineapples, they smell sort of sweet and sicky. Tomatoes smell sort of sweaty, like armpits.'

We paused at a mushroom wholesaler's stand. I asked the man in charge: 'How would you describe the smell of mushrooms?' He was emphatic: 'There are many mushroom smells. There isn't one basic smell.' He whipped out a jar containing black truffles and gave us a sniff. Nick reeled back, saying 'It's gassy. It's horrid.' The mushroom man went on: 'There are a whole different set of smells when you cook them. A lot of fungi smell really meaty when cooked. That's why they are such an asset to chefs.'

In another alley we found a wholesaler of cheeses. Like mushrooms, cheeses offer a palette of smells which give full play to the talent of the human brain for discriminating between subtle variations in odour. I was fortunate to find that the young man at the cheese stall was French, as French people are often better-informed and less inhibited than English people about aroma. Our conversation was however interrupted at intervals by Nick, who is firmly opposed to cheese (except when it takes the form of pizzas, gratins, gougères, cheese sauces or pasta toppings). I asked the young French man about his favourite cheeses. 'Well, it depends how ripe they are. Personally, I love Livarot. It reminds me of the countryside.' Nick butted in: 'Excuse me – I would like to say something – cheese – it's not my thing – by the way.' Undeterred, the French man continued: 'Roquefort smells to me of wet cellar. All the blue cheeses have a particular tang.' Nick chimed in: 'Sweaty – blue cheese smells sweaty and sicky – is that right?'

I spoke up for English cheeses, which when not factory-produced, mature into strong, piquant flavours. We sniffed a mature Lancashire cheese, and agreed that it smelled fruity, with a hint of peach or apricot, and a peppery sharpness. Nick made his contribution: 'It smells tense – it smells out of tune – like Stravinsky.' The washed cheeses next caught our attention. 'Is it true,' I wanted to know, 'that the aroma of Epoisses is so powerful that it has been banned from public transport in France?' My young French man insisted that this was a myth, and referred to advice from the official '*Syndicat de Défense de l'Epoisses*' to the effect that the cheese's smell could readily be masked on public transport when wrapped in damp newspaper or cabbage leaves. In his view, Epoisses was one of the glories of French civilisation, and it would be unpatriotic to ban it. Nick did not agree: 'It's horrid; it smells like someone has been sick.' So what, we asked Nick, do you like if you do not like the smell of good cheese? 'I like the smell of bacon,' he replied.

So we stepped out of the sales area, and looked for one of the greasy spoon cafes to be found in New Covent Garden. Dawn was staining the sky. The smell was of Full English Breakfast. I remembered a quote from J.B. Priestley: 'We plan, we toil, we suffer – in the hope of what? … Simply to wake up just in time to smell coffee and bacon and eggs … What a morning, what a delight.'

Later, Nick and I talked again about the smell of three o'clock in the morning. What, I asked him, does he think of three o'clock now? 'It's quite quiet,' he said, 'except for the big markets.' And the smell, I asked? 'It smells of dreams – the dreams I've had – BAD ones. There was a fox – I used to get up in my cot and wait for the fox – it was coming – it was coming to get me. What are foxes like – are they quite …? What's the nose of a fox like – pretty powerful, isn't it? In my cot, I stood up for the fox. It breathed.'

So, although our trip to the big wholesale markets had left me with a vivid memory, for Nick it was overshadowed by the distant memory of a bad dream when he was two years old. For a small child blind from birth, the terrors of the night were embodied, not in shadows behind the wardrobe or scary shapes under the bed, but in a breath and a smell. Such is the power of olfactory memories, and such is the vividness of Nick's memory (he claims to remember the hospital where he was born), that the terrors of the night could be revived by one whiff.

☙

There is a recurring dream that some people experience at intervals. It is a dream about an unopened door. In the dream, you are in a familiar house, and notice an extra door that you have not noticed before. You open the door, and suddenly you are in a large, sunlit room, one that you never knew existed, often with huge windows giving on to unexpected views. You realise that this house which you knew so well has unopened doors and secret spaces, bigger and brighter than the familiar spaces. The dream, I am sure, is a metaphor for our own minds, which have unopened doors and unexpected spaces.

The discovery that the sense of smell has hidden meanings was, for me, very like the dream about the unopened door. I had noticed the sense of smell, and I had wondered what was behind it, what meanings it conveyed. But I had not opened the door, until I set about exploring this sense. Having opened the door, I find that there is a large space behind it with unexpected views.

Most people would dispense with the sense of smell more readily than their other senses. It does nothing vital for the essential activities of life. Without it, we can still feed ourselves, work, sleep, move, navigate, interact with other people normally. If we lost it overnight, our family and friends might briefly sympathise, but they would rapidly forget that we had suffered a loss. There are no visible social consequences to being without the sense. Large numbers of people lose their sense of smell in old age, and a very few are born without one at all. Those without this sense do not immediately qualify for medical attention, special training or social security, as they would if they lost their sense of sight. They can find no high street shops offering olfactory tests and smelling aids with designer frames.

Yet the sense of smell supplies us unobtrusively with a constant stream of information about our environment. The smells of houses tell us of homely and unhomely places, of hygienic or unkempt dwellings, of cooking and ironing smells. The smell of the street gives us a whiff of each shop as we pass, of creosoted fences, of mown lawns and clipped hedges, of uncleared rubbish, of the nationality of fast-food shops. The smell of the countryside gives us dank leaves, lingering smoke, the crusted muck of animals and farm machinery, of passing brambles and honeysuckle.

The smell of people tells us about our loved ones and our kin, about the emotional states of people, about their smoking or drinking habits, occasionally about their state of health. It gives us vital information about our own smells, without which we might be in a state of social unease. Most important of all, perhaps, the sense of smell gives us a flow of information about food and drink. Without smell, most of our food would be just cardboard and most of our drink would be just liquid. We would miss all the subtlety of cooked flavours or of fermented drink, and much of the subtlety of raw food. We would miss some really important information such as whether food is tainted and has started to putrefy.

All this is the daily stuff of smelling, the breath-by-breath background information which we do not need to strain at. But we could, if we wish, open the door a little wider and find the unexpected space behind it. To open the door, all that is required is to pay attention to what we smell, to get in the habit of identifying what we smell and (when we are in luck) to put the smell into words. The door opens only a little at first, and we obtain only an obscure glimpse of the space behind. But our olfactory system is in a constant state of regeneration, and we will grow more smell neurones if we take an interest in what they tell us. The sense of smell improves with practice – not just because we become sharper but also because we expand those parts of the brain that deal with olfaction and the vocabulary of smells that they hold.

It is easy to suppose that we pay little attention to smells because our brains are just not equipped to do so. It is easy to suppose that our brains have developed in ways that are incompatible with understanding smells, and that other creatures which habitually pay close attention to smells have brains that are quite different. But that is not the case. It turns out that, although there are creatures capable of amazing feats of olfaction in limited domains, we humans are not duffers. We are reasonable all-rounders at smelling, and we have one amazing olfactory talent that other creatures do not seem to possess – the talent of creating flavour using our sense of smell in combination with other senses. It looks likely that the reason why we give a low place to our sense of smell is not because it is defective, but because we need to leave plenty of space in our brains for thinking.

It could be that the human sense of smell is very far from being semi-redundant, a left-over from the millennia when our ancestors inhabited the jungle, a pale shadow of the sense that dogs and rats use so well. Rather, it could be that our sense of smell is strikingly well-adapted to our needs. We are social creatures, and we have a sense of smell that gives us plenty of useful social information. We are home-making creatures, and our sense of smell tells us a great deal about the homely and the unhomely in our environment. We are creatures that cook, and our sense of smell gives us all the subtlety of flavour that results from cooking. We are omnivorous, and our sense of smell gives us a huge variety of smells and a powerful means of discriminating amongst them. We are multi-sensory creatures, and our sense of smell integrates brilliantly with our other senses. It also acts as a subtle amplifier of other important brain activities, enhancing our response to food and drink, modulating our emotions, sparking our inquisitiveness, even spurring our creativity.

The link between creativity and odour has hardly been touched on so far. Schiller, the German poet whose *Ode to Joy* set to music by Beethoven is now the European anthem, relied on a particular odour for his creative impulse. Goethe described visiting Schiller at his home and finding that his friend was out. Goethe decided to write some notes while waiting for his return. He sat at Schiller's desk; a peculiar smell prompted him to pause and to open the drawer in the desk from which the smell originated. Inside the drawer he found a pile of rotten apples. The smell was so overpowering that Goethe became light-headed. He had to go to the window for a breath of fresh air. Later he asked Schiller's wife Charlotte about it. She told him that Schiller had deliberately let the apples rot. The aroma inspired him and, according to Charlotte 'he could not live or work without it'.

Old apples for Schiller, madeleines for Proust, lemon-scented soap for James Joyce, these are little keys to a creative door in their brains. Once the key is turned and the door opens, there are big spaces behind and unexpected views. One unexpected view is of the variety of some quite ordinary things. The smell of wood, for instance, seems a simple enough category – there is the smell of wood, and we recognise it in woody perfumes and woody wines. But in reality, the variety of wood aromas is wider than we can imagine – larch wood is meaty; cherry wood is fruity; lime wood smells of babies; English oak and bog oak and Turkey oak each smell different; sandalwood smells beautiful and iroko wood smells repulsive; and so on through all hardwoods and softwoods and tropical woods. There is no single wood smell (it is probably the high resinous smell of pine that most of us have in mind as the smell of wood), but a wide variety of smells. The smell of vinegar, to take another instance, is simple and distinctive. It is distinctive enough that most of us could probably summon up a rough idea of it in our heads: sour, fruity, aggressive, and the essential aroma of fish and chip shops. But once we put our noses to the variety of vinegar smells, we find that the field is vastly varied and nuanced. There is real balsamic vinegar, and faux balsamic vinegar (the kind we are most likely to encounter in the supermarket – wine vinegar with some flavouring, not the sort that is matured for years), rice vinegar, cider vinegar, vinegars made from beer, sugarcane, coconut, honey, sherry and any number of fruits, vinegars flavoured with any number of herbs or spices. There is a long list of other substances like wood and vinegar, which seem limited in odour at first sniff, but prove to have limitless nuances – cheeses, tobacco, wine, petrol, geraniums, apples, grass matting, coffee, tea, mints, basil, fermented meats and many more.

The sense of smell, like the dream of the unopened door, opens up new perspectives on the world. One such perspective is the view that odours offer us into the nature of things. The molecules that convey smell come from the inner substance of the things that smell. It is a closer contact with the substance of things than our other senses afford – sight gives us superficial qualities like shape and colour, sound gives us the vibration of air around things; but smell gives us volatile fragments of the thing itself. The act of perceiving a smell therefore puts us in direct contact with the thing, and also with other things that share the same volatile constituents and other things that have undergone the same chemical changes. A whiff of frying bacon, one of the glories of the English breakfast, tells us of the plate we are about to eat, but also it tells us of other things that share the same molecules of smoke, of charring, of pork, of hot oil.

Through the sense of smell, a door opens on quite another room in our brains – our personal past. Many of the smells we recognise are childhood smells, long preserved intact in our brains, and when we open the door on the sense of smell we allow ourselves to travel backwards in time. Swept away by a smell from our youth encountered much later in life, we are able to recall the whole sensation of childhood – its smell, but also the look, the feel, the emotions of that distant time. Memories sparked by vision or hearing do not work in the same way; we are often left feeling puzzled and detached by an old snapshot or recording of our childhood. Smell memories are altogether more powerful.

Opening the door into our brains that is represented by our sense of smell brings us to a space with yet another surprising characteristic. The sense of smell amplifies our other senses and our emotions, and the space it occupies in our brains acts as something of an echo chamber. This effect is most obvious in flavour – the smells that we savour in food and drink enhance and round out the sensations that are conveyed by our taste buds, our sense of feel and the other senses. Without smell, food and drink are very incomplete experiences. It is not simply that we have lost the nuances supplied by our noses; it is that we have lost the amplification of all the senses that smell provides. By the same token, an experience in which we engage our sense of smell is a bigger experience, and gives more pleasure. It is not simply that an extra sense has been added, but that all the usual senses are enhanced by the sense of smell. So the experience of a walk in the country or a trip in the city is amplified when we consciously give a place to smelling. All the senses are brighter as a result, and we live more in the present. That attribute of being more alert to the present moment is, in part, what Rilke meant when he wrote of savouring some sprigs of heather, and remarked that, without giving attention to all our senses 'one comes into the present unfinished, unable, distracted'.

This amplification also works for the emotions. The sense of smell can change our emotional state, and emotions can change our perception of smells. The two-way exchange between olfaction and emotions is special, and is not found with the other senses. It is explained no doubt by the fact that our sense of smell operates within the brain on very close terms with our emotions. It accounts for the link between smells and the first and most powerful of our emotions – the bond between baby and mother. When someone has been separated from his mother for a long time and is finally reunited, mother and child will bury their noses in each other's necks, and let the sense of smell remake the bond.

I found myself idly watching the TV news one day, and two items of news in close succession illustrated this theme. One featured a refugee from the Syrian war searching for his long-lost mother. The camera crew followed him as he encountered a convoy of buses carrying displaced people from his home town. In one bus was his mother. Their meeting was highly emotional and wordless, and the camera lingered on it. He said just one thing to his mother – 'Let me smell you.' The other news item featured a cheery astronaut, returning to earth from months in a space station. Asked how he felt, he struggled for words, and said 'The smell of the earth – you know you are back.' In both stories, it was the sense of smell that carried the bond between man and mother.

CHAPTER 17

Afterword

'It has been observed that one's nose is never so happy as when it is thrust into affairs of another, from which some physiologists have drawn the inference that the nose is devoid of the sense of smell.'

Ambrose Bierce – *The Devil's Dictionary*

In 1960 Steffen Arctander, a Danish scientist of fragrances based in the US, published a great bible called *Perfume and Flavor Materials of Natural Origin*. The book covers in 700 pages every source of fragrance from Abies Alba oil (silver spruce) to Zdravetz oil. In his preface to the book, Arctander wrote that he found himself in August 1958 in possession of an almost complete collection of perfume and flavour materials, and became aware that there was no work which described such materials in everyday words, nor any work which gave practical indications of use and availability. He aimed to fill those gaps, and he succeeded magnificently.

It is a book to idle through, travelling the exotic trade routes of little-known fragrances, distant fields and plantations, tiny distilleries of essential oils, and obscure exporters. Arctander illustrated the book with his own photos of jasmine extraction in Morocco, clove buds in Madagascar, eucalyptus in the Congo, vetiver grass in Réunion, lemon blossom in Cyprus, and elderflowers in Denmark. Each entry gives you a quick pen picture of the material, its provenance, its smell, and how it performs as a perfume. Here, for instance, is the entry for Zdravetz oil:

> Zdravetz oil is produced only on demand. The plant *Geranium Macrorhyzum* is a small perennial which grows on rocky soil at high altitudes in the Balkans and South-east Europe. The oil is distilled in Bulgaria … The odour of Zdravetz oil is sweet-woody with a floral and faintly herbaceous undertone, reminiscent of clary sage, tobacco, broom absolute, tea leaves and *ulex europaeus*. The floral notes are of rosy-woody character, delicate and tenacious … The oil has an excellent fixative effect which can be utilised in fougères, chypres, crêpe de Chine ….

That short description summons up for me a small crowd of characters from the 1960s – Balkan women in head scarves wandering the rocky hillsides to harvest the plant, Bulgarian distillers in ill-fitting Communist era suits haggling over the price of a batch of Zdravetz oil, elegant perfumers in laboratories juggling the ingredients of a chypre perfume, and a diligent Dane recording the facts about the material.

Sadly this great thesaurus has been largely left behind by the march of commerce. Materials of natural origin have largely been superseded by synthetics. Arctander himself produced a companion volume called *Perfume and Flavor Materials of Chemical Origin* in 1969. His book on natural materials, written just nine years before, had come at the very end of an era. Only aromatherapists hang on to his original book about natural materials, and see it as the bible of their craft.

I would like to see a thesaurus like Arctander's covering, not perfume materials, but all the smells of everyday life. Its purpose would be to record descriptions of smells accurate enough to summon up the sensation in my head, and to communicate the smell to others. In extreme old age, when I will be too frail to get out to savour the smells directly, I will leaf through the thesaurus, remembering the sensations that the smells gave and the associations they evoke. I conceive of this work attracting entries from contributors around the world, who will, by adding their own descriptions, and polishing those of others, gradually build up a widely-accepted thesaurus of hundreds of common smells.

It is a tall order to describe smells in a way that others will recognise and accept. The perception of smells is a very subjective activity, with many people perceiving the same smell in slightly different ways, with markedly different associations and preferences. Two experienced perfumers, even two perfumers who have worked in the same laboratory for years, are unlikely to give identical descriptions of the same material or the same perfume. For them, that is part of the romance of perfumery. Steffen Arctander had this to say about the problem:

> The description of the odour of a perfume has been, and still is, the source of endless discussions among perfumers, pseudo-perfumers and laymen. The more exactly one attempts to describe a material, the fewer people are able to agree with the author on his description. Some brief and basic very general terms from everyday talk and work would seem to offer the best midway solution to the problem. Obviously no odour can be described verbally in any language in such a way

> that every reader will immediately visualise the material and be enabled immediately to identify it if he is faced with an unlabelled sample of the material ... No descriptions are unambiguous or even very striking. But if we could describe every single one of our perfume materials in such a way that no two descriptions were alike, and so that every one fitted like a key to only one material, we could lean back on a wreath of laurels. We would have conceived the impossible: we would have invented the perfect and foolproof odour classification system.

With that admission that his descriptions would be imperfect, he went on to give a succinct description of the odour of every one of the perfume materials he had collected. Where he wanted to describe something he did not possess, he insisted on going to the source and refused to rely on secondhand accounts. Following in the footsteps of Arctander, I offer my descriptions of over 200 ordinary smells, making use like Arctander of 'some brief and basic very general terms from everyday talk'. I have tried to use the four-sniff approach advocated in this book to discern how complex or simple a smell is; what related smells it resembles or differs from; how bitter or sweet it is; and what can be said about its piquancy or other attributes. Not surprisingly, some smells took a good deal longer to unravel; some because they are too complex to be easily unpicked, some because they are too simple for words. I am sure that these descriptions are imperfect, and hope that the reader will be impelled to improve them.

These are the rules I set myself at the outset:

- Each entry should have a single line at the start which captures the distinctive character of that smell. After that first line, there can be any number of supplementary notes about nuances.
- Each entry should only employ words that are in common usage, and make references only to substances likely to be encountered in everyday life.
- No personal likes, dislikes or associations should feature.
- There should be no reference to the source of the smell, nor to its uses (to avoid circular descriptions such as 'almond smells almondish', or 'almonds smell of marzipan').

An A–Z of Smells

Almond (essence) Concentrated, creamy nut; very sweet like vanilla with a woody tang.

Almond (ground) Mild, bready, creamy nut; sweet but with a catch-at-the-back-of-the-throat that is common to most nuts.

Ammonia Shocking, breath-taking; too pungent to smell of much, other than a hint of seaweed.

Anchovy Rich, concentrated, oily stink of dried fish; sharp and metallic; quite peppery. Affinity with strong mature cheese and strong cured meat.

Angostura bitters Rich, multi-layered aroma, dominated by warm mixed spices and treacle; with a strong element of sweet citrus, and an unusual flower scent (gentian); faded alcoholic headiness.

Anise (star anise) Rich liquorice; earthy, dark and treacly, with a tarry edge; combined with a dusty spiciness like dry tobacco.

Apple Gentle, creamy, homely fruit; sweet with a touch of sharpness; woody with a hint of tannin; hint of sour milk. Many variants around the main theme.

Apricot (dried) Gentle, warm, jammy sweet; halfway between prune and peach; creamy.

Artemisia Complex blend of cat, lavender and rose. Major flavouring for vermouth.

Asparagus Fresh, a strong version of grass, with an edge of nuts like green hazelnuts.

Bacon (grilled) Very strong, very savoury, like a salty jam.

Balsamic vinegar Complex, combining sour vinegar with smokey sweetness like demerara sugar; hint of herb like basil; piquant. Affinity with coal tar.

Banana Creamy, sweet aromatic tropical fruit, with afternote of old gym shoe.

Basil (fresh) Complex redolent herb; sharp like tomato stem; zesty like lemon; oily like clove.

Bay leaf Complex savoury herb, combining citrus, cured meat and leather.

Beer (bitter) Watery, malty, with a faintly bitter herbal smell; reminiscent of damp plaster.

Beer (lager) Metallic, malty; hint of honey; hint of bitter medicinal flower; faint alcoholic fumes.

Beer (stout) Malty, dark like a watery chocolate, caramel, with an edge of bitterness.

Beetroot Earthy, sweet, fruity; like a blend of parsnips and port wine.

Birch (log) Soggy newspaper smell; faintly sour.

Blackcurrant Rounded, rich berry; gummy sweet with a medicinal or sulphurous edge.

Bonfire Very complex; sweet smoke carrying wafts of wood, grass, vegetable, animal, even fish; very acrid close up, but soft at a distance.

Book (glossy) Light wood and china clay, with a chemical edge like Humbrol paint.

Book (paperback) Old soggy wood with a hint of vanilla and a faintly surgical smell like methylated spirits.

Box leaf Homely blend of mown grass and old wood, with a faint smell of spiced fruit (like spiced apple puree).

Bracken Warm grass, with strong notes of fig and fennel which emerge slowly.

Brandy Rich, suave, complex fermented raisin; strongly alcoholic; with touches of vanilla and oak.

Bread (rye) Rich, dark, malty; akin to stout or stewed fruit; with a bitter edge.

Bread (white, French) Simple and fresh, combining warm malty smell with a hint of milky acidity; reminiscent of sawdust.

Bread (wholemeal) Mild, comfortable, malty smell; hint of a floury fruit like baked apple or apricot; hint of sour milk.

Brown sugar Rich, dry sweet, with a hint of smokiness like pipe tobacco.

Brussels sprout Cabbagey with an undertone of leek; hint of roasted chestnut.

Butter Mild, creamy, slightly rancid; reminiscent of the smell of a new baby; hint of maize; slightly metallic.

Cabbage Slightly offensive smell of green stuff, sulphurous, bitter, reminiscent of sweat.

Campari Bitter-sweet; fruity; resinous; with a hint of gentian.

Candle Waxy, resinous, honeyed, ecclesiastical.

Capers Fresh, nutty, peppery, slightly metallic; background smell of vinegar; mildly piquant.

Car (new – interior) Artificial-smelling combination of sandalwood, berry and plastic.

Caraway seed Earthy anise; nutty and slightly tarry; akin to fennel but stronger and more nutty.

Cardamom Very cool, bitter-sweet spice; slightly oily and bitter like cloves; freshness akin to citrus fruit.

Cardboard Damp wood chippings with glue, and a faint medicinal edge.

Carrot Fresh, simple and mild; earthy; faintly sweet; faintly woody.

Cauliflower Cabbagey, but creamier and more sweet than cabbage.

Cedar Warm and resinous; the smell of pencils.

Celery Watery, bitter, electric smell; with an afternote of soy.

Cheese (blue) Strong, sweaty, rancid smell; close to bacterial smell of old socks; background of acid like lemon; faintly piquant.

Cheese (Munster – runny) Powerful, ripe and foecal; buttery; like long-unwashed body.

Cheese (Parmesan) Complex mix of creaminess, citrus and a raucous savoury quality close to sweaty feet.

Cherry Very faint smell (unlike its taste); the simplest berry smell, balancing sweet and tart.

Chocolate Rich, vanilla-sweet, nutty; thick, roasted smell like a sweet gravy; suave.

Cigar Rich, cloying, fragrant smoke, with many layers of grassy, tarry, earthy, spicy, bitter and sweet notes.

Cigarette Light fragrant smoke, sweet with a tarry, bitter afternote.

Cinnamon Warm, sweet, fruity spice; with a thick sweetness that combines raisins and ginger, and a dry freshness that has hints of citrus and apple.

Cistus Delicate, smokey incense, with a touch of sharpness like rosemary.

Cling film Mild plastic, highly reminiscent of chicken fat.

Clove Rich, heady and complex spice; a blend of sweetness (as in apple) with oily bitterness (as in burning plastic), but with a power all its own.

Coconut Sweet, buttery, fragrant nut.

Coffee (roast) Multi-layered, rich, dark, burnt, nutty. Like dark chocolate, but with a bitter edge like burnt toast.

Coir matting Dry, dusty, with the sweet/sour tannin smell of oak bark.

Cola Blend of sweetened lemonade, cinnamon and burnt caramel.

Compost Dungy; a complex mix of sour, gassy smells; rich fruitcake smells; and fungal, earthy smells.

Coriander (ground) Warm, citrussy spice, like a blend of orange and sage; hint of turpentine.

Courgette Barely perceptible vegetable smell, like a watery version of cucumber, with a faint sulphur tang of seaside.

Crab Sweet, chalky shellfish; with a sharp, gassy edge; the smell of the seashore.

Creosote Very rich, resinous, old; smokey like tar; animal note of leather; bitter edge like burnt sugar.

Cucumber Fresh, mealy and cool; a blend of bread, grass and cool herb.

Cumin (ground) Cool, fresh, smokey spice; with a vegetable freshness akin to vinegar; gently piquant; affinity to potatoes and tomatoes.

Cypress Mild resinous blend of turpentine and citrus.

Dill Cool and sweet herb, with a touch of vinegar and a very slight piquancy.

Dog Dank animal smell that combines laundry (hot and damp) with seaside beach (chalky, marine).

Eucalyptus Complex, multi-layered smell, mainly a warm minty resin, with hints of apple and citrus.

Fennel seed Sweet, rich, dusty herbal smell, blending liquorice/aniseed with dry tobacco.

Fenugreek Rich, curry-like; a blend of celery and maple syrup; piquant.

Fig (dried) Rich, old blend of stewed fruit and leather; sweet like plum, but heftier, with a slight edge of burnt caramel. Affinity with rye bread.

Fig leaf Blend of young bracken and stewed fruit; sweet and fragrant, with a slight edge of cat's pee.

Fir (wood of Douglas fir) Fresh, light, highly resinous.

Flour (white) Faint, dry, nutty, chalky.

Fruit cake Complex rich smell of stewed fruit and fermented malt; sweet with a faint bitter edge of burnt caramel.

Fuel oil Rich, suffocating mix of bad egg (sulphur) and fermented fruit.

Garlic Gassy, raucous, fresh oniony smell, aggressive and clinging. Afternote of old cheese like Parmesan.

Geranium First whiff is astringent and mineral, like a pebble beach, followed by the normal, grassy smell of leaf.

Gin Delicate, fruity alcohol; oily; with hint of juniper berry, spicy and leathery.

Ginger (fresh) Cool, sweet blend of honey and lime; highly piquant.

Glue (neoprene) Aggressive, like a strong white wine combined with methylated spirits; underlying smell of wax or polish.

Goat's cheese Rancid like old socks; faintly citrous; faintly honeyed.

Grapefruit Most powerful of citrus smells; very fresh and sharp; richer and sweeter than lemon, fresher and sharper than orange.

Green pepper Very fresh and loud vegetable smell, akin to broad beans or runner beans, but with a rubbery, metallic edge.

Ham Strong dungy, flatulent smell; hefty like wholemeal bread; hints of honey and of smoke.

Hay Sweet dry, aromatic grass, with many touches of flowers, especially fennel and mint.

Hazelnut Faint, green, oily nut; halfway between a green twig and a pistachio nut.

Hemp rope Rich, old blend of hay, leather and fermented meat, with an edge like old sweat.

Honey Pungently sweet, waxy, hefty like bread; hint of citrus.

Horse Comfortable blend of some uncomfortable smells – dung, urine, hay, dust.

Incense Resinous, sweet and citrus smoke; lingering; exalted.

Ivy Sour, flat, green smell; with a hint of citrus.

Jasmine Creamy sweet; the simplest, purest sweetness amongst flowers.

Jet fuel Cloying petrol smell, with notes of mint and tropical fruit.

Juniper berry Rich, woody spice combining the smells of cypress, leather and berry.

Jute Rich, leathery and grassy; like a blend of hay and wet dog.

Kiwi fruit Initial smell of foul gas, followed by a mild berry smell like unripe raspberry; hint of sharp citrus and vomit.

Larch wood Meaty resin, like salami and turpentine.

Laurel Mild blend of green bananas and almond.

Lavender Cool incense, sharp-sweet, with an enveloping assertiveness. Hints of eucalyptus, cistus and geranium; but a power all its own.

Leaf mould Old, earthy, dark; the smell of woodland in winter.

Leather Old, rich, animal smell with oaky tannin.

Leek Mild version of onion, with a hint of sweetness and of ripe cheese.

Lemon Fresh and simple; the archetype of citrus; refreshingly bitter but with an underlying sweetness as well.

Lentils A complex, hefty, farinaceous smell, close to cooked meat and to stout, with a musty, earthy feel; mealy, meaty, malty.

Lettuce Fresh and simple; creamy version of grass with a slight bitter edge; hint of peanut butter.

Lime (wood) Creamy curdled smell; reminiscent of babies.

Linseed oil Warm, oily blend of fresh-cut grass, nut, beeswax and treacle.

Lubricating oil Simple, faint smell of liquid metal with a hint of rubber.

Mace Like nutmeg, a rich, sweet, nutty spice; slightly more piquant and savoury than nutmeg.

Mango Very sweet, smooth and juicy; with an intensity close to pineapple or strawberry.

Maple syrup Malty, oily treacle, with a hint of coffee.

Marjoram Cool, sweet herb; akin to apple; akin to green tea; close to oregano but more woody; faint medicinal edge.

Marmalade Sharp, dark, cooked orange, jammy sweet with a tart edge.

Matches Light woody smoke, with the fishiness of sulphur.

Melon (Canteloupe) Warm, sweet, fragrant, watery; hint of goat's cheese as it ripens; hint of petrol.

Methylated Spirits Sour and astringent medical smell, aggressively pungent.

Milk (fresh) Very faint smell; a touch of cheese, a touch of chalk; a touch of edgy sourness.

Milk (off) Edgily rancid; halfway between gentle milk and old socks; metallic touch of bitterness.

Mint Cool, almost icy herb, combined with fresh green apple. Evolves from initial chilliness to a fragrance close to sage. Many varied scents in different species of mint.

Mushroom Dry, mouldy smell as in leaf mould, with a hint of oily fish.

Mustard (Dijon) Aggressive, bright, vinegary, vegetable smell.

Mustard (English) Aggressive like pepper; eggy; bitter like vinegar.

Nail varnish Intrusive, aggressively sweet, slightly foecal; a blend of the smells of glue, rubber and banana.

Nasturtium Sweet and peppery, close to the smell of pea and of cress.

Nectarine Warm, juicy, woody; cooler than a peach; warmer and more fragrant than a plum.

Newspaper Faint soggy wood, with a chemical edge of printers' ink.

Nutmeg Rich, sweet, nutty spice; hint of citrus; hint of medicinal oiliness; hint of smoke.

Oak (wood) Gentle tannin, with a touch of sour milk and of fruity vinegar.

Oats Sweet, mealy smell; honeyed; with a nutty quality like hazelnut.

Olive oil Fresh, fruity oil; concentrated greenness like crushed grass; nutty afternote.

Onion Gassy and fresh; halfway between offensively sulphurous and mildly sweet; makes eyes water.

Orange Fresh, sweet and simple; sweeter and darker than lemon, but with the characteristic sharp zest of citrus.

Oregano Cool, smokey, lemony herb; close to marjoram but warmer.

Paint (undercoat) Rubbery and petrol with a sweet mint undertone.

Parsley Very fresh, simple, powerful herb; like essence of green grass.

Pea Simple, fresh, sweet, mealy like beans.

Peach Warm, fragrant, sweet fruit; halfway between apricot and plum; faint afternote of rubber.

Peanut Earthy nut with a chemical hint of polish.

Pear Gentle, watery, sweet like mild version of apple, with cheesy edge and hint of animal like wet wool.

Pepper (black) Highly piquant – provokes a sneeze; dry and resinous with an affinity to incense.

Petrol Light, astringent and aromatic mineral smell. Hints of citrus, bad egg and perfume.

Pine Strident, very resinous, light, intrusive.

Pineapple Sugary-sweet, fresh and watery, tropical in intensity; hint of sour milk.

Play-Doh Strong almond essence, overlaying floury, chalky smell.

Plum Fragrant, juicy and woody; halfway between blackberry and peach.

Polish (beeswax) Powerful mix of honey and white spirit; touch of incense; pleasantly pungent.

Port Plummy, gummy wine with a dark, coffee-like undertone.

Potato (uncooked) Earthy and fresh vegetable; faintly nut-like; akin to crushed grass; touch of leaf mould.

Prawns (cooked) Pronounced smell of boiled egg, with only a hint of shellfish.

Privet Sweetish leafy smell that has affinities with damp wrapping paper and vanilla.

Putty Mealy like uncooked cake mix, with strong scent of linseed.

Radish Faint, watery, musty; hint of vegetable sweetness like carrot and sour leaf mould.

Raisin Rich, sweet, old fruit; almost a fermented or stewed smell akin to port or stewed prunes.

Raspberry Sweet and sharp berry, with homely undertone of bread.

Redcurrant Very fresh berry; close to wild strawberry but more acid; faint and simple.

Redcurrant leaf Startling blend of cat's piss and berry.

Rice (brown, cooked) Creamy, grainy; like porridge with a hint of cocoa.

Rhubarb Astringent and fruity; earthy; a sharp, fruity version of celery.

Rose (bourbon) Sweet, powdery and smooth, with a touch of coconut. Reminiscent of suntan lotion.

Rosemary Rich blend of incense, dry citrus and grass. Ecclesiastical and sharp at the same time.

Rubber Loud, cloying smell all of its own; mix of gassy, fruity, resinous and fishy.

Rue Shocking mix of ammonia, honey and cat.

Rum Sweet, smokey treacle and fermented fruit, with a powerful alcoholic headiness.

Sage Strong aromatic herb, combining a meaty savouriness with a bitter edge of cats' piss; mildly piquant. Many varieties with highly distinctive smells.

Salami Rich, fermented stink of pork and garlic; close to cured ham; close to leather; hint of smokiness.

Salmon (Cooked) Suave and creamy, sweetish, chalky smell close to smell of shellfish.

Salmon (Uncooked) Sharp, metallic, fishy smell.

Seagrass Close to hay but more fragrant: dry, grassy, hint of liquorice.

Sesame oil Savoury nut, tarry and grainy; like concentrated brown toast.

Shrimp (brown) Sharp, gassy smell of seaweed comes first, followed by succulent, chalky sweetness common to shellfish like prawns or lobster.

Soot Faint, sour, resinous smell of tar and damp ash.

Strawberry Very sweet and fragrant, like a blend of vanilla and ripe melon; creamy, with a hint of tainted milk or even vomit.

Sweetcorn Sweetest of vegetables; like a blend of maple syrup and cauliflower; faint hint of fermentation; faint hint of bleach.

Sweet potato Smell of damp potato combined with burnt caramel; reminiscent of burnt top of apple crumble.

Tangerine Mildest of citrus fruits; sweet and zesty; comfortable marmalade smell; evocative of Christmas.

Tar Rich and thick; a blend of chocolate, bad egg and petrol.

Tarragon Dry smokey herb; pronounced dark touch of liquorice; undertone of leather.

Tea (Indian) Old, earthy and woody, like woodland undergrowth; metallic with a hint of bitter tar.

Tea (Lapsang) Complex blend of smokey, floral and fruity; smokey as in tarry ash; floral as in cistus; fruity as in dried fruit with a citrus edge; old like a wooden ship.

Tennis ball A blend of soft rubber with gas; afternote of fish paste.

Thyme Complex dry smokey herb; warm and tobacco-like with hints of sweetness and citrus; faintly piquant.

Toast Malty with a hint of walnut; comfortable with a bitter edge of burntness.

Tomato Fresh, watery vegetable; both sweet and savoury; with a sharp herbal tang like basil. A fresh air smell with a touch of the seashore about it.

Tomato ketchup Complex, rich smell of overripe, gummy tomato with mild vinegar; hints of clove, sweat and garlic.

Treacle (golden syrup) Simple, honeyed; as smooth as butter; hefty like bread.

Truffle oil Strong, complex, gassy; more powerful than other fungal smells, with touches of earth, metal and strong cooked meat.

Tuna Oily, sweet, metallic fish; like a blend of vanilla and old shellfish.

Turmeric (ground) Complex balance of warm and cool; at the same time musky, earthy and citrous; like a blend of orange, roast vegetables and ginger.

Turpentine Startlingly resinous and astringent. Verging on the sharpness of lemon.

Vanilla Extremely sweet, rich and creamy; with a deep tropical sweetness; akin to bananas but more redolent.

Varnish Redolent resin with alcoholic piquancy; sour hint of lubricating oil.

Verbena (herb tea) A soggy blend of lemon drops and damp leaf.

Vinegar Startlingly sour grape with burnt caramel.

Watercress Simple, sharp and peppery; reluctant to yield its distinctive smell until crushed or chewed.

WD40 Vanilla-flavoured version of lubricating oil.

Whisky Delicate smell of grain and smoke; heady, almost medicinal.

Wine (red) Rich, smokey, fermented grape; with a dark sweetness as in stewed fruit and a slight bitter edge of vinegar; gentle alcoholic piquancy. Many different nuances for different wines.

White spirit Light, berry-like version of petrol. Mildly aggressive and clinging. Reminiscent of polish.

Whiteboard marker Powerful blend of alcoholic and medical smells, like a blend of brandy with meths.

Yogurt Inoffensively sour milk; creamy and citrussy; hint of vomit and of honey.

Notes

1. Darwin in *The Descent of Man* wrote with apparent authority about the sense of smell, but with evidence that now looks shaky. He asserted that mammals have highly developed senses of smell for identifying prey and danger, but that in humans the sense is of extremely slight service. He quoted second-hand evidence that has not stood the test of time to the effect that Africans and Indians have more developed senses of smell than 'white and civilised' races, and concluded that the sense of smell is disappearing as human beings evolve.

2. Freud had a close interest in the sense of smell, influenced by a Berlin ear, nose and throat surgeon called Wilhelm Fliess. Fliess treated Freud for some persistent nasal problems, administering cocaine and operating on him several times. Freud developed a theory of smell which centred on the idea of repression. He extended his ideas to a general theory about society which asserted that the more civilised a society, the more it sublimates the sense of smell. As with Darwin, Freud's ideas about the sense of smell involve hypotheses about evolution which depend on crude notions of racial superiority and civilisation.

3. The 1994 book *Aroma* surveys a number of different cultures with strong 'osmologies': classificatory systems based on smell. These include the cultures of the Desana and Suya Indians of Brazil, the Andaman Islanders and the Sever Ndut of Senegal.

4. Joy Bear and Richard Thomas, 1964. That many dry clays and soils produce a characteristic odour when breathed on or moistened with water had been recognised for a long time. Such an odour was widely associated with the first rains after a period of drought, and there was evidence that drought-stricken cattle respond in a restless way to this smell of rain. The smell had been successfully captured by a small perfume industry in India, and fixed with sandalwood oil in

a perfume called '*matti-ka attar*' or 'earth perfume'. Bear and Thomas extracted a yellowish oil by steam distillation from dry rocks, and demonstrated that it was responsible for the odour.

5. '*Natural Sniffing Gives Optimum Odour Perception for Humans*' – David Laing, 1983. Laing tested people's natural sniffing techniques against other techniques where the number of sniffs, the interval between sniffs, and the size of sniffs were varied. He concluded that humans achieve optimum odour perception with their natural sniff, with an average inhalation rate of 30 $1 min^{-1}$, volume 200 cm^3, and duration 0.4 s.

6. The 'tip-of-the-nose' effect was labelled by the psychologists Lawless and Engen in 1977. It is akin to the experience of recognising a face but not being able to put a name to the face, which is known as the 'tip-of-the-tongue' effect. Memories for faces and for odours have much in common. Like faces, odours give social signals, elicit instant emotions, are difficult to describe, are perceived as a single image not as a set of details, elicit context-dependent memories and are quite resistant to forgetting.

7. '*The Two Smells of Touched or Pickled Iron (Skin) Carbonyl-Hydrocarbons and Organophosphines*' – Dietrich and Glindemann. In this paper, Dietrich and Glindemann explain the source of the 'metallic' odour that is generated when someone handles keys, coins or metal objects. The odour results from a metal-induced oxidation of skin lipids. The compounds in the smell contain no iron. 'The fact that iron metal produces a whole host of smelly organic molecules when humans touch it or acid attacks it was unexpected. While our deciphering of the chemistry is scientifically fascinating, it also has a wide range of benefits, from designing tests to monitor human diseases associated with oxidative damage to cells, to improving blood-scent tracking.'

8. Charles Sell, who spent most of his career as the leading research chemist for one of the six major 'fragrance houses', wrote an article in 2006 for a specialist chemistry journal entitled '*On the Unpredictability of Odor*'. He wrote that, despite a great deal of research on the relationship between the structure of a molecule and its smell, it is still not possible to predict with certainty what a new molecule will smell of. It is possible to establish by statistical methods the *probable* character, threshold and intensity. But it is only a probability, and there are often surprises. Serendipity continues to be an important factor in the discovery of

novel fragrant molecules. He concluded that it is the nature of olfaction that the smells of molecules are inherently unpredictable.

9. In 1949 Moncrieff proposed what became the mainstream theory of how nerve receptors in the nose detect smell molecules. His 'shape' theory proposed that the molecule's shape fits like a key in to olfactory receptors. This has been modified by the 'odotype' or 'Weak Shape Theory' which proposes that receptors recognise only small structural features of each molecule, and that the brain identifies the combination of activated receptors as a smell. The 'vibration' theory is proposed by L Turin (1996). The molecular volume theory was proposed by Saberi and Seyed-allaei in 2016.

10. Olfactory ensheathing cells (OECs) are unique cells that are responsible for the successful regeneration of olfactory axons throughout the life of mammals. They are the only nerve cells known to regenerate. A painstaking programme of research at the Institute of Neurology at University College London explored, first with paralysed animals, how OECs could be transplanted from the nose into injured spinal cords to stimulate the regrowth of nerve cells there. Their studies attracted worldwide interest and were applied to paraplegic human patients at Wroclaw Medical University in Poland. A patient there, Darek Fidyka, whose spinal cord had been severed in a stabbing, had one of his olfactory bulbs removed, the OECs grown in a culture, and then implanted into his severed spinal cord. A year and a half later, his spinal nerve fibres had regrown across the gap sufficiently for him to have recovered some voluntary movement and sensation in his legs, to walk and to drive.

11. A number of research projects have confirmed that pregnant women experience changes in their sense of smell after about 36 weeks of pregnancy. Many experience strong aversions to particular smells, a few experience strong impulses to sniff particular smells. It is hypothesised that these changes serve a useful purpose in steering women away from foods that could damage their babies, and encouraging women to broaden their diet to supply nutrients needed by the baby. Testing their olfactory function shows that they do not change their sensitivity to smells or their ability to distinguish smells, but rather that their smell preferences change during pregnancy.

12. Richard Axel and Linda Buck won the Nobel Prize for Physiology in 2004 for their discovery of 'odorant receptors and the organisation of the olfactory

system'. They discovered a large gene family comprising some 1,000 different genes (3 per cent of our genes) that give rise to an equivalent number of olfactory receptor types. Their fundamental paper was published in 1991.

13. Koelega and Koster, in research published in 1974, showed that threshold sensitivities to smells – the lowest concentration of a smell that can be detected – can vary by a factor of 1,000 from person to person.

14. '*Odor signatures and kin recognition*' – Porter, Cernoch and Balogh, 1985. Humans can detect by body odour people who are blood-related kin (mothers/fathers and biological children, but not husbands or wives). '*Possible olfaction-based mechanisms in human kin recognition and inbreeding avoidance*'. Weisfeld, Czilli, Phillips, Gall, Lichtman, 2003. Mothers can identify by body odour their biological children but not their step-children. Pre-adolescent children can detect by smell their full siblings but not half-siblings or step-siblings.

15. *Making Rounds With Oscar: The Extraordinary Gift of an Ordinary Cat* – Dr David Dosa, 2010. As of January 2010, Oscar had accurately predicted approximately 50 patients' deaths. Dr Dosa commented that certain chemicals are released when someone is dying, and that Oscar is able to smell them. Feline experts have observed that cats can often sense when their owners are sick or when another animal is sick. Some believe that cats can sense when the weather will change, and when an earthquake is imminent. There may be other ways, however, in which cats can detect these things, for instance from movement or lack of movement.

16. '*Ministry compiles list of nation's 100 best-smelling spots*' – *Japan Times*, 31 Oct 2001. 'In a bid to focus attention on pleasant smells rather than malodorous ones, the Environment Ministry has decided to showcase 100 aromatic sites … The selected smells cover everything from Chinese medicine shops in Toyama Prefecture and hedges in Aomori to seaweed shops in Mie and grilled eel in Shizuoka … The most heavily debated sites were Tsuruhashi Station in Osaka which was nominated for the smell of barbecued meat and kimchi, and Tokyo's Kanda area which was nominated for the scent of bookstores.'

17. *Very long-term memory for odors* – Goldman and Seamon 1992. *Odor Memory: Taking Stock* – Schab 1991. Research on smell memory confirms the everyday impression that such memories are stable over very long periods, and that the curve for forgetting them is different to memories laid down by other senses.

18. A landmark paper by the psychologist Jonathan Haidt at the University of Virginia in 2001 proposed that instinctive gut feelings, rather than logical reasoning, govern our judgements of right and wrong. Haidt and colleagues demonstrated that a subliminal sense of disgust – induced by hypnosis – increased the severity of people's moral judgements about shoplifting and political bribery, for example. A variety of experiments, particularly by David Pizarro of Cornell University and Yoel Inbar of Tilburg University in the Netherlands have shown that disgusting ambient smells affect the moral judgments of people of all political complexions, tending to make their judgments more severe.

19. For newborn mammals, the first vital test of life is to find their mother's nipple, usually in competition with siblings. In most cases blind and deaf at birth, they rely on touch and smell. There are many smell cues for the newborn to follow – the mother's milk, secretions from glands around the nipple, smells transferred as a result of the mother licking the newborn and the nipple area. The most complete understanding of newborn behaviour stems from the rabbit. Baby rabbits show a typical pattern of searching in the mother's abdominal fur by smell. When the smell of the rabbit's milk was analysed by gas chromatograph, one molecule (2-metyl-but-2-enal) emerged as the likely pheromone eliciting the baby rabbit's behaviour (Schaal et al 2003). Human babies also use smell cues to orient to their mother's breasts, especially from the milk and from Montgomery glands around the areola. As with other species, there is an intricate host of smells. So far, no evidence has emerged for a smell stimulus that would qualify as a pheromone.

20. Wedekind and colleagues at the University of Bern conducted research in 1995 to establish whether humans have preferences, as mice do, linked by smell to immunotype. Male volunteers slept for two nights in the same T-shirt, and female volunteers sniffed the T-shirts. Women preferred those worn by men whose immune system genes did not match their own. However, women taking the contraceptive pill preferred the odours of men with similar genes to their own. Wedekind surmised that the balance of hormones circulating in the blood of women taking the pill is similar to that of pregnant women, and that during pregnancy women are influenced by different odour preferences. Subsequent research on large populations of humans has tended to confirm that there is a statistical tendency for mate choice to correlate with MHC genes in the way found by Wedekind, but it is not a firm rule, and is modified by a variety of other factors.

21. *'Bidirectionality, mediation, and moderation of metaphorical effects: the embodiment of social suspicion and fishy smells'* – Lee and Schwarz 2012.

22. Experiments by Hasler and Wisby in 1954 with coho salmon in the USA demonstrated the role of smell for salmon homing on their native stream. One experiment involved plugging the noses of some salmon; those salmon whose noses were plugged missed a crucial turning in the stream which would have taken them home, while the other fish did not. In another experiment, Hasler exposed smolts (young fish undergoing physiological changes that prepare them for migration) to a chemical to see if they would later home to a river containing that chemical. Hasler exposed smolting coho salmon in Wisconsin hatcheries to the chemical morpholine, and then transported them to Lake Michigan. The fish did not have home rivers to return to, so Hasler simulated these by putting morpholine into some of the rivers flowing into the lake. The coho, identified by fin clips, homed to those rivers by the thousands.

23. Sobel and Porter, University of California, Berkeley, 2006. The experiment was designed to test not only whether humans can follow a scent trail, but also whether we can locate smells by comparing the differences between the two nostrils, in the same way that we use the differences between our two ears to locate a sound. The research established that, although our nostrils are close together, they sample smells from areas sufficiently far apart to allow us to distinguish the edge of a scent plume, and so locate a smell.

24. Martial was a Roman poet writing in the century after Christ. Born in Hispania (modern Spain), he is best known for his books of epigrams; short, witty poems in which he satirises city life and the scandalous activities of his acquaintances and romanticises his provincial upbringing. His epigrams still shine with wit and realism, making more recent satirists seem laboured. He captures smells, good and bad ones, with extraordinary sharpness. Here, for instance, is Martial on a kiss: 'Breath of a young maid as she bites an apple; perfume such as when the blossoming vine blooms with early clusters; the scent of grass which a sheep has just cropped; the odour of myrtle, of the Arab spice-gatherer, of rubbed amber; of a fire made pale with Eastern frankincense; of the earth lightly sprinkled with summer rain.'

25. *'The Cool Scent of Power: Effects of Ambient Scent on Consumer Preferences and Choice Behaviour.'* – American Marketing Association, 2014. Cinnamon and

peppermint scents improve cognitive performance (Zoladz and Raudenbush, 2005). Clerical work in offices performed better when exposed to the smell of peppermint (Barker et al, 2003).

26. '*The muted sense: neurocognitive limitations of olfactory language*' – Olofsson and Gottfried 2015.

27. At a scientific conference in 2010, asparagus was served at dinner and the scientists' conversation turned to the bitter smell of urine after eating asparagus. Some scientists were familiar with the sulphurous tang; others had no idea. The experience prompted Prof. L Mucci and colleagues at Harvard to investigate, contacting 6,900 people to establish how many could smell the peculiar tang of urine after asparagus, and whether inability to smell it was because the smell was absent or because it was not perceived: 58 per cent of men and 61.5 per cent of women reported that they could not smell the odour. The reason proved to be genetic abnormalities in the sense of smell – hundreds of variations in DNA were found which linked to the inability to perceive the smell.

28. Frantext is an online collection of works of French literature at the University of Lorraine. A search of Frantext for the use of terms referring to smells and to colours by major French novelists of the 18th, 19th and 20th centuries showed a surge in the use of smell terms by 19th century writers. On this crude measure, Zola comes out top, though this tells us nothing about how well he uses olfactory language.

Author	**No. of novels on Frantext**	**Smells**	**Colours**
Diderot	5	5	39
Rousseau	1	5	14
Stendhal	7	30	247
Balzac	87	232	2576
Hugo	6	72	811
Flaubert	7	396	770
Verne	11	36	493
Zola	22	1034	2943
Proust	7	189	549
Aragon	4	51	582
Céline	3	108	344

Acknowledgements

I would like to acknowledge, first, the writers who have shown that it is possible to put the sense of smell into words – Marcel Proust, James Joyce, Rainer Maria Rilke, Aldous Huxley and (in a quite different vein) Steffen Arctander.

Second, I would like to acknowledge those on whose scientific knowledge I have relied – Professor Thomas Hummel; Professor Richard Axel; the late Oliver Sacks; the authors of 'Olfaction, Taste and Cognition', but especially Egon Peter Köster; Luca Turin; the authors of the historical and anthropological survey 'Aroma'; and the many contributors to Wikipedia.

Third, I acknowledge with gratitude Tom Webber of Icon Books who took a risk on this unpromising project and kneaded it into shape, and Humfrey Hunter, my agent. My brother Stephen has been part of the literary team, helping with his experience in the world of books and his moral support.

Finally, I want to acknowledge the love and patience of my immediate family – Toll with his enthusiastic support; Nick with his unflinching interest in smells; and wonderful, loving Frances.

PERMISSIONS

- The extract from *Right Ho, Jeeves* by P.G. Wodehouse on p 1 is reproduced by permission of the estate of P.G. Wodehouse c/o Rogers, Coleridge & White Ltd, 20 Powis Mews, London W11 1JN.
- The extract from *Flush* by Virginia Woolf on p 113 is reproduced by permission of Houghton Mifflin, Copyright 1933 by Houghton Mifflin Harcourt Publishing Company. All rights reserved.
- The extract from *The Road to Wigan Pier* by George Orwell on p 165 is reproduced by permission of Bill Hamilton as the Literary Executor of the Estate of the Late Sonia Brownell Orwell.
- The panel from Calvin and Hobbes on p 203 is © 1995 Watterson. Reprinted with permission of ANDREWS McMEEL SYNDICATION. All rights reserved.
- The extract from W. Somerset Maugham's *On a Chinese Screen* on p 223 is copyright the Royal Literary Fund.
- The letter printed on p 249 is from *Letters on Cézanne* by Rainer Maria Rilke – edited by Clara Rilke, translated by Joel Agee, published by Jonathan Cape and reprinted by permission of The Random House Group Limited.

Translations from the French, where appropriate, are the author's own.

Every effort has been made to contact the copyright holders of the material reproduced in this book. If any have been inadvertently overlooked, the publisher will be pleased to make acknowledgement on future editions if notified.

ABOUT THE AUTHOR

Barney Shaw is an artist and former civil servant. He was Private Secretary to several Labour and Conservative Ministers, and went on to run government policy on various aspects of work, unemployment and schools. This is his first book.